Integration of Methods Improvement and Measurement into Industrial Engineering Functions

Integration of Methods Improvement and Measurement into Industrial Engineering Functions

Gerald J. Watson Jr.
with illustrations by Jesse J. Derouin

CRC Press
Taylor & Francis Group
Boca Raton London New York

CRC Press is an imprint of the
Taylor & Francis Group, an **informa** business

First edition published 2022
by CRC Press
6000 Broken Sound Parkway NW, Suite 300, Boca Raton, FL 33487-2742

and by CRC Press
2 Park Square, Milton Park, Abingdon, Oxon, OX14 4RN

© 2022 Gerald J. Watson Jr. and Jesse Derouin

CRC Press is an imprint of Taylor & Francis Group, LLC

Library of Congress Cataloging-in-Publication Data

Names: Watson, Gerald J., Jr., author. | Derouin, Jesse J., illustrator.
Title: Integration of methods improvement and measurement into industrial engineering functions / Gerald J. Watson ; with illustrations by Jesse J. Derouin.
Description: First edition. | Boca Raton : CRC Press, [2022] | Includes bibliographical references and index.
Identifiers: LCCN 2021025351 (print) | LCCN 2021025352 (ebook) | ISBN 9780367720933 (hbk) | ISBN 9780367720940 (pbk) | ISBN 9781003153412 (ebk)
Subjects: LCSH: Operations research. | Continuous improvement process. | Project management. | Work measurement.
Classification: LCC T57.6 .W38 2022 (print) | LCC T57.6 (ebook) | DDC 658.4/034--dc23
LC record available at https://lccn.loc.gov/2021025351
LC ebook record available at https://lccn.loc.gov/2021025352

ISBN: 978-0-367-72093-3 (hbk)
ISBN: 978-0-367-72094-0 (pbk)
ISBN: 978-1-003-15341-2 (ebk)

DOI: 10.1201/9781003153412

Typeset in Times
by Deanta Global Publishing Services, Chennai, India

Dedication

This book is dedicated to my wife who helped me get out of the Georgia Institute of Technology, not once but twice. She then provided moral and financial support when I returned at age 59 to work toward a PhD in Industrial and Systems Engineering at the North Carolina Agricultural and Technical University in Greensboro and earned it at the age of 64.

It is also dedicated to the many students over the years that have helped me become a better person and instructor.

Contents

Foreword

As CEO of Everest Group International (EGI), I have a need for accomplished consultants with an extensive experience matrix uniquely blended with strong academic credentials and original thought who can unveil hidden opportunities and forge new directions. Everest Group International provides advisory services to private equity groups, Boards, and company owners by focusing on transitions and transformations. EGI achieves paradigm-shifting results by advancing operations optimization, accelerating earnings, and increasing a company's valuation.

To the delight of our clients, Dr Watson personifies our ideal consultant profile. Someone who leads our healthy client companies to front-runner status. Someone who not only helps a company achieve a successful turnaround but then relaunches it to sustainable profitability. One attribute in particular is important to EGI. Dr Watson quickly assesses a company's operations – even those that appear first class – and discovers paths to consequential improvement.

As one might expect, not all clients are receptive to change. Interestingly, the more resistant, the weaker the bottom line. In these cases, Dr Watson embraces the manufacturing team and invites them to identify problems and opportunities, join in the leadership of the project, and share ideas for improvement. In such instances, he coalesces the team into a unified resource and mobilizes the involved employees.

Dr Watson has come to the rescue of EGI on innumerable occasions including some that are quite memorable. He has an uncommon skill set, not the least of which is his expertise in impactfully applying his methods improvements and measurement strategies to live-action situations. He enlightened me to the fact that methods improvements and measurements can frequently lead to the discovery of untapped upsides.

In the EGI world, any engagement has the potential to turn into an adventure. I recall such a situation that involved a Texas-based furniture manufacturer with national distribution that utilized a captive trucking firm. EGI was brought in by the private equity firm who owned the company to address productivity issues throughout the organization. This turned out to be the least of their problems. It was a classic story of money, guns, and lawyers. We discovered a drug cartel had infiltrated the company, including middle and upper management, then utilized it as part of its supply chain for moving product from Mexico to Texas, and subsequently smuggling drugs in company shipments for distribution nationwide. As you might imagine, there was no better person to invite into this situation than Dr Watson. We faced havoc and chaos as we went about improving the company's performance while simultaneously restructuring the management group. Despite the circumstances, Dr Watson was able to optimize, combine, and eliminate steps in the manufacturing process while streamlining warehouse operations and maximizing truckload capacity within the constraints of proper shipping methods. A happy ending to a rocky start.

Sometimes the impossible must be made possible. I'm reminded of a situation where a client who manufactured consumer hard goods had the opportunity to open a new big-box retailer by going directly to a large nationwide test. To do so required many impossibilities. CNC equipment had to be identified, sourced, installed, and made operational. Custom packaging had to be designed and client approved. The manufacturing operation itself had to be staffed to run 24/7 and all the operators had to be trained and quality assurance criteria met. Dr Watson stepped forward to play a crucial role in making all this happen.

The application of methods improvements and measurements can be wide ranging. I have a memory of a company that manufactured custom-designed and engineered metal buildings. Estimating the cost of projects for customers was complex, time consuming, and fraught with error. Dr Watson studied this process and developed a costing tool using multivariate statistics with an accuracy rating of 95%. He first segregated projects by level of complexity and developed a new methodology for high frequency, recurring building types. This was just one of many improvement and productivity initiatives. Based on transformations attained in conjunction with the management team, the company was subsequently sold to a large U.S. steel producer at a premium price.

I remember another engagement with a national manufacturer and distributor of windows and doors. At the time, the company was owned by a private equity group that was in search of a quick but substantial impact on earnings. Dr Watson studied, analyzed, and changed methods, practices, layouts, and flow to realize an optimization level that amazed and gratified management and factory employees alike. Many elements were involved including value stream mapping. Another outcome of this assignment was increased inventory turns not only for Work-in-Process but also raw materials.

It's clear that Dr Watson's comprehensive approach including emphasizing methods improvements and measurements produces results that have a direct and immediate impact on the P&L. This, in turn, increases the value of the enterprise which is an objective of all the stakeholders. So, while Dr Watson's techniques for methods improvements and measurements produce gains in the various operating areas, they also translate into success in the boardroom.

EGI has been the beneficiary of Dr Watson's efforts on many occasions. Some hard, some tough, most of them fun, and a few close to impossible. Reading this book will give you further insight into his approach and secrets of success. You will have an entirely new appreciation of how methods improvements and measurement can play a key role in a company's success, financial strength, and viability.

Larry D. Hughes
CEO
Everest Group International

Preface

The reason that this book was written is to share my experiences, successes and failures, and the lessons learned from both successes and failures to increase the readers' likelihood of success in all endeavors. During the many years that I worked as an industrial engineer, manager, and consultant with different companies in diverse industries, I committed many mistakes for a variety of reasons that provided learning opportunities to prevent a repetition of the same mistake.

The several factors that make this book unique are the presentation of situations that if present in your organization demand that resources be allocated to address them to enable the organization to continuous improvement and long-term survival. Based on the varied experiences of the author solutions for each situation will be proposed. Another factor contributing to the uniqueness of this book is the presentation of both successful and unsuccessful examples of the various types of improvement projects on which that industrial engineers will focus.

1. Every work environment afforded me the opportunity to work with experienced and knowledgeable people. Those who have been in a position for several years have more knowledge about that operation than someone who recently graduated from college regardless of the college and the number of degrees. Based on my thirty years' experience, most people will share that information if asked appropriately.

2. It is better to complete a project late than on time and fail. As a former football coach at Ga Tech always said, "if you are not first, you are last". Over the years our football team has been last many times. But no one is hired for the failures on the field but for their numerous successes over the years. The failure as a project leader will be remembered forever but projects that were successful but late will soon be forgotten.

   Before the initiation of a project and during each step, always ask what could go wrong. Anticipate and plan for unanticipated mistakes. Murphy's law will always appear at the worst possible time.

3. Take responsibility for your mistakes and ask for help to mitigate the effects and advice on the prevention of similar future mistakes.

   A plant manager provided a quote for a large potential customer. The quoted price was 20% less than if priced using the price list. As the sales to this company increased, the profitability of this plant declined. It was obvious that the prices quoted were inadequate to cover the direct cost of producing and shipping the product.

   At a meeting held at the plant to discuss options, prior to the start of the meeting, the plant manager stood up and told those in attendance that he had really made a mistake and would appreciate the assistance of everyone to address the situation to prevent the mistake from shutting the plant down resulting in the loss of many jobs. Those in attendance included the president of the company, the chief financial officer, mechanical and industrial engineers, and management from other company plants.

After his announcement, the attitude of everyone changed and all were dedicated to improving the situation. Since the quoted prices were guaranteed for a year, the only solution was to reduce the costs of this and all other products produced at this plant. The group immediately established a proactive plan and assigned responsibilities for each aspect of the plan. At that time the company did not have a standard cost system which was the first item on the agenda. This task was assigned to industrial engineering. The task of improving quality levels was given to quality assurance. The task of increasing automation to reduce direct labor costs was assigned to the mechanical engineers.

Follow-up meetings were held weekly. Within six months a standard cost system was in place, quality improved as a result of better training for all employees and documented specifications, and conveyors were installed which reduced direct labor costs.

The result was an overall costs reduction sufficient to offset the losses incurred due to the low price which enabled the plant to become profitable. Costs of the item on which the low price was submitted were reduced significantly. The standard cost of the item was calculated with the newly established standard cost system and was equal to the price quoted. This team effort resulted in the conversion of this product from losing money to a break-even situation.

Another benefit was the standard cost system which would prevent a similar situation in the future.

4. Never assume or take anything for granted. Always check and double check or measure twice but cut once. Assumptions can result in the loss of lives, money, time, and trust.

An example of the result of assuming and not checking as needed resulted in the loss of a spacecraft that was designed to study the atmosphere on Mars. The spacecraft was lost due to the assumption that data supplied by the contractor was units in the metric system, which was standard practice for space missions. The contractor provided data in the U.S. imperial system. The navigation engineers at another location assumed incorrectly that conversions had occurred. The contractor supplied information using the English System (pound). The cost of the orbiter spacecraft in 1999 was 125 million USD [1]. This amount of money would have paid for many checks and double checks.

To improve one must first have a metric to indicate the current position as productivity percent, cost of quality, and accident and injury rates. This current position should be a static one, like being stuck in traffic, but the goal is forward movement or improvement. Based on the many years of experience of the author, many tools and techniques will be presented in detail that will result in improvement. Equally important are standardized evaluative techniques to be used after improvement has occurred to insure that improvement did occur and that the savings from the improvements exceed the costs of these improvements.

The motivation for writing this book is to follow up a previously written one that emphasizes the use of learning from past mistakes and critical thinking to develop a strategy to prevent future ones. The author has over thirty years' experience of making and witnessing the mistakes made of experienced people who developed a plan that they thought would result in a successful ending. However well planned and implemented, the project did not produce the expected results. This book was written to provide a guide that if properly followed, will result in the success of the projects with which industrial engineers are involved.

Information about various topics that will assist the industrial engineer to achieve improvement goals will be discussed to demonstrate how this information can be applied in the improvement phase to achieve the goals. For example, ergonomics will be discussed and applied to the design of the workplace that can increase productivity and reduce accidents and injuries. Usability will be discussed to demonstrate how its use will result in products that are easier to be used and instructions easier to follow.

This book is unique due to the incorporation of the experience of the author and his goal which is to increase the effectiveness of those in the field of Industrial Engineering. A previous text titled "Critical Thinking-Learning from Mistakes and How to Prevent Them" was focused on the application of critical thinking as a learning tool to prevent future mistakes. The contents of this book will address issues that a decision-maker will confront initially to prevent the commission of a mistake.

The contents are applicable to students, new graduates, and practitioners to increase their job productivity and longevity. The book uses examples from the author's experiences and those he has witnessed to demonstrate the importance of critical thinking in the application of industrial engineering tools in the workplace. It achieves this to insure that at least the following occur:

1. All reasonable possibilities of solutions are taken into consideration.

   Various techniques, as barnstorming, are available to insure that this occurs. Employees with experience are crucial to insure that previous mistakes are not repeated as are those who propose new and innovative solutions.

2. All costs are included in all projects.

   Often costs that may be unknown at the time needed to make a decision due to factors that will be determined more accurately after the decision is implemented. Based on the experience and opinion of the author, all costs should be known in advance of requesting the change. For those costs over which the requestor has no control, as negotiated freight rates, a cost that should be at the upper reasonable limit should be added since it is better to realize results that exceed expectations than for results to be less than expected.

   All costs include incoming freight, packaging, inspection, handling, waste, etc. must be included to enable cost comparisons to occur.

3. Projects are completed on time and within budget.

   All projects regardless of scope or budget must have a fully developed and detailed project plan. The responsible person must insure all tasks are completed as scheduled to insure that the critical path is maintained. Also critical is timely feedback to enable adjustments to be made as needed to insure that deviations from the critical path do not occur.

4. The consequences of Murphy's law are considered.

   Based on the experience of the author and discussions with numerous people in various professions, this law cannot be ignored and solutions must be developed in advance due to the inevitability of the occurrence of this law.

5. That continual improvements become the normal mode of operation in this organization. These improvements may occur in productivity improvements, the working environment, the reduction of accidents and injuries, increased morale of employees, improved quality, an increase in customer satisfaction, etc.

6. That accident, incident rates, and near misses are continually decreased.

   The environment in which employees work must constantly be evaluated to reduce the risks of injuries and accidents. This goal must have top management support that is responsive to an active safety committee.

7. That manufacturing throughput is continually improved.

   The result will be improved delivery and service which is achievable as a result of a coordinated effort involving all areas of the organization.

8. That raw material and work-in-process inventory levels are reduced as much as feasible. Inventory quantity, including work-in-process, should be continually monitored to obtain the correct balance between too much and too little. This balance will depend on numerous factors, including the time to replenish the inventory, the cost of the item, the cost of not being able to manufacture due to a lack of inventory, the probability of obsolesce and damage during storage and handling, the amount of warehouse space occupied, as well as the cost of storage and retrieval of the inventory.

   The author, based on his experience, would prefer having to explain an excess quantity of inventory than to explain to a customer that the delivery will be delayed due to a lack of inventory particularly if the item that resulted in the delayed shipment was relatively inexpensive. His experience has actually witnessed this situation which resulted when a truck supplying foam from a long-term vendor ran out of fuel en-route to the plant resulting in an interruption of production scheduled for that day, missed and late deliveries and in the use of overtime to enable the plant to return to its normal production schedule. The vendor made several changes to improve service, the most significant was a maintenance check to insure that all vehicles in its fleet were available and capable of making all deliveries.

9. That the company operate as environmentally friendly as possible.

   The company should incorporate this philosophy throughout its operations from purchasing to delivery and shipment. All means of reducing its environmental footprint need to be fully investigated.

10. That the material used for the packaging of a product for shipment is appropriate for its mode of delivery.

    The result of this effort will include a reduction in packaging costs and a decrease in the environmental footprint.

11. That transportation costs have been analyzed to maximize opportunities for cost reduction as a result of increasing the quantity of items per shipment and increased usage of backhauling and similar services.

    The success of this analysis will reduce transportation costs and the environmental impact of the company but will also improve delivery and customer satisfaction.

12. That overhead costs have been evaluated to determine possible methods for reduction.

An example is the shifting of processes of production from a shift with high utility costs to one with lower utility costs. A company that manufactured paper board from recyclable materials that experienced high utility costs incurred from drying the better grades of paperboard on first shift began producing the higher quality of board on third shift due to lower utility costs during off peak. The higher quality of board required a slower drying speed and increased supervision and inspection which was the primary reason it had been produced on the first shift. The cost savings in the cost of electricity due to shifting production from the first shift to the third exceeded the additional costs of supervision and quality inspection that were now required on the third shift.

Another example of utility cost savings resulted as a company that produced fiberglass insulation shifted its drying operation from the first to the third shift. The costs of drying the wet fiberglass during off-peak usage more than compensated for the additional supervision and inspection that was needed on the third shift.

The book achieves these goals by the presentation of examples of the application of methods improvement and measurement to the various challenges that industrial engineering will confront. These examples will include both successful and unsuccessful ones. The critical questions that were asked that resulted in success will be discussed in detail. The unasked critical questions that resulted in failure will also be presented and discussed in detail. The opportunities for improvement that will be presented include the establishment or the modification of existing direct labor standards, or the modification or the installation of an individual or group incentive system. The evaluation of an existing or proposed process change, providing assistance in the selection of a new vendor, or the evaluation of a current one, and the installation of new or the evaluation of a safety program.

The first chapter titled "Methods Improvements" provides a definition of a method and a detailed explanation of the process of method improvement but it also enumerates many situations based on the experience of the author that are opportunities for potential improvement. Reasons for the improvement opportunities are provided as well as proposed solutions. Each of these proposed solutions is fully explained in a chapter in the text.

Critical thinking will be used to demonstrate that its use will result in the successful integration into the desired industrial application by the use of actual examples and the presentation of the questions that were asked prior to beginning the implementation to increase the likelihood of success. A failed application of that same tool and the questions that should have been asked to insure its success to corroborate the need for critical thinking will also be discussed.

Acknowledgments

Dr Watson wants to recognize the assistance of Mr Larry Hughes, who provided valuable guidance and input during the writing of this manuscript. Mr Hughes, also an alumnus of the Georgia Institute of Technology, has successfully managed with ever-increasing levels of responsibility numerous companies during his corporate career and as a consultant has worked successfully to enable companies to remain viable and increase profitability. He has also provided numerous opportunities for Dr Watson to gain additional experience and knowledge.

Contributor biographies

Gerald J. Watson Jr. has worked for over 30 years in manufacturing as a manufacturing engineer/manager and safety team leader in various industries with different companies both as a full-time employee and as a consultant. During that period, he gained valuable insight into the attitudes and the factors that motivate a majority of employees regardless of industry, company, and position or job title with the company.

Jesse Derouin has worked for over ten years accumulating valuable industry experience. Jesse has had the benefit of having worked in several capacities gaining an understanding of how to successfully combine people, equipment, and processes to create favorable outcomes. As an adult returning student, he set about achieving an undergraduate degree in engineering, where he met Dr Watson. He is now a graduate student studying for his advanced engineering degree in sustainable system design.

Glossary

Item	Definition
Calculation of the incentive rate	The incentive rate is calculated by multiplying the standard hours earned with the units produced which yields standard hours earned for a particular operator. This is then divided by the number of hours the operator worked on the incentive plan which provides the incentive rate.
	It is the ratio of standard hours earned by the number of standard hours worked while working on the incentive system
	It is the product of the number of standard hours earned and the operator's base rate of pay.
Definition of an operator rating of 100%	Normal performance is the rate of output which qualified workers will achieve without overexertion over the working day shifts provided they know and adhere to the specified method and provided they are motivated to apply themselves to the work. This performance is denoted as 100% on standard rating and performance scales.
Methodology of a formal time study	Introduce yourself to the selected operator and request permission to be studied.
	Explain to the operator the need for the study and how it will be conducted.
	Use as many cycles as needed to obtain detailed elemental descriptions.
	Begin the actual recording of times using the start and stop times from the elemental descriptions.
	Record all fixtures and tools used by the operator.
	Ask the operator questions during the study to insure that all necessary tasks are included.
	Note any time the operator is idle and the reason, e.g., no parts, making repairs.
	Insure the recording of enough cycles to obtain the accuracy level.
	Calculate the average time for each element.
	Calculate the mean and standard deviation of the average times.
	Verify that the desired accuracy level is obtained – if not obtain additional cycle times.
	Sum the average times – this is the normal time for that operation.
	Multiply the normal time for that task by the rating given to the operator.
	Multiply by the allowance percentage – the result of the standard time for that operation.
	Convert the standard time to hours using three decimals.

Introduction

Definition of Industrial engineering
Brief history
Tools and methods of Industrial Engineering
Applications and results

The official definition of Industrial Engineering as defined by the Institute of Industrial and Systems Engineering, (IISE) is as follows:

> Industrial and systems engineering is concerned with the design, improvement, and installation of integrated systems of people, materials, information, equipment and energy. It draws upon specialized knowledge and skill in the mathematical, physical, and social sciences together with the principles and methods of engineering analysis and design, to specify, predict, and evaluate the results to be obtained from such systems. [2]

The application of industrial engineering has effected all aspects of our lives. Since its initial role in the standardization of manufacturing processes and the establishment of production standards, applications have expanded into many other areas, as logistics and healthcare. The goal of these current methods is to determine how these current methods can be improved as a result of the use of ergonomics and usability. The foot-activated trash can that eliminates the need to bend over to raise the top is the result of the application of ergonomics [3]. The application of Taylors' four principles to manufacturing has resulted in improved product quality, reduced costs, improved customer service, and a greater availability of products [4].

Utility testing has increased the ease with which products, services, and especially communications can be used. A pair of scissors is an excellent example because its use is intuitive. An example of increased usability in communications is the smartphone, which provides a trained user with numerous options like the ability to communicate with others, conduct banking transactions, and obtain directions to the next destination. Examples of increased usability in the automobile are an indicator on the dashboard that points to the side of the automobile to use to refuel and the installation of a camera that enables the operator to operate the vehicle in reverse successfully [5].

Manufacturing has benefitted immensely since it was in this arena that Scientific Management has its origins. Industrial engineering applications were soon adopted and further defined in the automobile industry by Ford Motor Company. From other automobile manufacturers, the field was quickly employed in other industries. In every application, various techniques were applied with the specific goal to increase productivity and efficiency [6].

DOI: 10.1201/9781003153412-101

Opportunities for improvement are basically unlimited. Results can be seen in the industries that vary and include banking, communication, healthcare, and transportation. From the traditional industries as manufacturing, industrial engineers today have begun to focus on different areas, such as banking, transportation and logistics, healthcare, retail, and non-profit organizations public service.

Some of the techniques include time studies, value analysis, workflow analysis, motion economy, lean manufacturing, and optimization techniques.

The application of industrial engineering techniques to healthcare has resulted in numerous benefits including a reduction of medication errors, nosocomial infections, and lengths of hospital stays. Improved communications concerning patient treatment and the scheduling of that treatment have resulted in improved patient satisfaction. Another area in healthcare that led to increased patient dissatisfaction is the length of time between various steps in treatment. The results of a study whose focus was to reduce the amount of time required for patients to receive treatment through the various steps during their hospital stay was the use of flowcharts [7].

One author focused his research for his dissertation to develop a format using human reliability and multivariate statistics to predict the percentage of errors and the severity of those errors in the various stages of medication administration. The study concluded that an overwhelming majority of errors (70.86%) was due to human error [8].

Examples of improvements in manufacturing have resulted in a greater variety of products and improved delivery. Many companies offer overnight delivery; some in as quickly as one day.

The modification of digital printers to print in three dimensions and the adaption of that technology to manufacturing have made an impact that will only increase. The adaption to manufacturing is referred to as additive manufacturing because this process adds material unlike other manufacturing processes that remove material. Additive manufacturing adds or prints one layer of material at a time starting at the bottom. This technology, used with 3D scanners and CAD (computer aided design software) creates or prints a product or image. This technology enables a manufacturer to generate items that meet specific customer needs. It allows for the creation of parts to meet specific customer specifications. Advantages include a reduction in lead times, a reduction of material costs resulting from less waste, and rapid prototyping. Ideal for rapid prototyping digital process enables the design alterations to be done quickly and efficiently during the manufacturing process. The lack of material wastage provides cost reduction for high-value parts, while additive manufacturing (AM) has also been shown to reduce lead times [9].

Applications of additive manufacturing or 3D printing include printing rapid prototypes for parts for cars and aircraft allowing manufacturers for testing of parts to insure that the parts will meet specifications before the need to allocate money for manufacturing the item on a larger scale. It is also used to insure to test the ability of the parts to will pass requirements before spending money on the large-scale manufacturing, and to build models to enable architects, engineers, and others to demonstrate plans to clients, saving time build resulting in saving costs and time.

The costs of this technology are decreasing and for some items are competitive with traditional manufacturing.

Applications have expanded into healthcare and include the development of patient-specific implantable devices as cardiac valves and potential source to repair or replace organs that no longer function as kidneys and hearts, or skin or any kind of organs. The material used to print these organs would be tissue from the patient reducing the likelihood of tissue rejection [10].

The many goals of industrial engineering focus on improving productivity and efficiency, reducing the number of defects and other wastes, reducing and the costs of injuries and accidents, improving product quality, reducing packaging costs, reducing waste and shipping costs, and evaluating overhead for potential reductions. Productivity is the ratio of standard hours earned to the total of actual hours worked and measures the relationship between output and one input, efficiency is the ratio of standard hours earned to the actual number of hours that the machine was productive. It is a measure of the productivity of the machine and measures the relationship between output and one input, usually measured in direct labor hours. The number of hours that the machine worked equals total time less time the machine was not running due to repairs.

Industrial engineers must consider the work environment of employees to insure that all employees can work productively and safely. To achieve this the engineer must take into consideration the human factor in relation to the technical aspect of the situation to reduce the discomfort of the employee. To achieve this goal, industrial engineers accumulate knowledge from numerous disciplines including human factors to improve the environmental conditions under which all employees work.

Based on the extensive experience of the author, it is the branch of engineering that is focused on continual improvement of any processes, systems, or organization by working with people at all levels to improve and implement integrated systems of needed resources, knowledge, information, equipment, energy, and materials. This is achieved through the development of training programs, the use of time studies for direct labor and indirect labor standards, and appropriate feedback for continuous improvement, adoption of effective safety programs, involvement of employees in decision-making to the maximum extent possible, development of quality inspection procedures, and other programs and procedures that can affect the productivity of the company and the safety of the employees.

Industrial engineers today can improve processes in any industry. Many make improvements in healthcare reducing medication errors and increasing throughput. Many work in the banking industry as can be seen in the improvements that are evident when making a deposit or withdrawal. Changes in the service logistics industry are obvious when companies offer same-day service for a minimum charge or one-day service for free.

Not only do industrial engineers work to increase the efficiency and profitability of their employers, but they are concerned with the environment through the study of many related fields, as anthropometry. For example, when designing a production line for maximum productivity, the measurements of those who will be working on the line must be considered as well as the weight of any items that must be lifted. When writing instructions, usability testing is to evaluate how well the instructions can be followed.

Industrial engineers must possess mathematical skills since they must be able to explain the results of time studies so that they can be understood. They also need to be able to explain a justification for a new piece of equipment or software or an upgrade. However, in the opinion of the author the greatest skill needed is people skills since the greatest idea in the world is not beneficial unless it is implemented and improved through the process of continual improvement.

One area that is essential in the design of the workplace is human factors. The Ames Research Laboratories of NASA defines it as "an umbrella term for several areas of research that include human performance, technology design and human computer interaction (HCI)" [11]. Focus at the Ames Research Laboratories is on research to improve the safety and efficiency of cost-effective operations both on the ground and in the air. Maintenance and training are an integral component of the research.

Ergonomics is not to be confused with human factors. The primary difference is that ergonomics focuses on the neck down whereas human factors include cognitive functions as well so it is concerned with the head down. Human factors include human physical, mental, and perceptual capabilities because all are required when studying work and improved work designs. Ergonomics focuses on adapting the workplace to the individual worker [12].

One such field is anthropometry, which is the science that defines physical measures of a person's size, form, and functional capacities. Applied to occupational injury prevention, anthropometric measurements are used to study the interaction of workers with tasks, tools, machines, vehicles, and personal protective equipment – particularly to determine the degree of protection against dangerous exposures, whether chronic or acute.

Anthropometry employs various static measurements of the human body to enable the human body to work productively with minimum discomfort. Another important aspect of anthropometry is the measurement of abilities as they relate to the movement involved in the performance of tasks, as reaching, motion and maneuvering, and other aspects in the use of various pieces of equipment. Both areas would be used by the industrial engineer in the design of workstations. [13]

Another field is usability. The term "usability" itself is focused on the ease with which the end user can operate or interact with a product or system. It focuses on the user, his satisfaction, efficiency, and effectiveness, which is whether or not applied, usability engineering can help reduce the number of mistakes that are committed daily by eliminating the opportunities for these errors or mistakes. As an example, automobiles have an indicator light on the dashboard to indicate on which side of the car the gas tank is located. This prevents the driver from attempting to pump gasoline into his car from the wrong side of the pump and becoming frustrated because the hose does not reach the tank. Another example is the use of green at gasoline stations on fuel pumps to denote diesel fuel. This is to notify the driver and to hopefully prevent the driver from adding the incorrect fuel to his vehicle resulting in costly engine damage. A reduction in medication errors occurred as a result of an anesthesiologist who, as he was adding fuel to his plane, realized that each type of fuel had a different color nozzle and could only be attached one way and applied that concept to the field of anesthesiology [14].

Usability testing is a method to improve a product or service by having users test a product or service. This is normally accomplished by assigning a group of people a list of tasks that one should be able to perform with the product or service without difficulty and then by determining how well the participants perform the assigned tasks. A pair of scissors is an example of a well-designed product since its use is intuitive to the user. The user can pick them up, understand the positioning of their fingers, and begin using the instrument for which it was intended intuitively.

If the product is software, then the time it takes to complete the task or the number of clicks on the mouse can be recorded for a metric. Some laboratories have eye-tracking software that measures the eye movements the eyes make in order to perform the assigned tasks. After the users have performed the tasks, the observers collect the results for analysis. Critical thinking must be used during this time to generate a list of questions that will improve the methods by which usability is improved. Measurement techniques will be developed to insure that continuous improvement follows each iteration [5].

A critical question that must be asked for every situation that requires a change in the data base is:

Who will initiate the Engineering Change Order (ECO) requesting the change and who is responsible for the accuracy of the information added to the data base?

This question is applicable in all situations, although the list of critical questions in each chapter may not specify this essential question. The function requesting the change is responsible for initiating the ECO. The reason is that the person performing this function is knowledgeable about the current situation and the effects of the proposed change. This change could be a cost reduction, a change in the specification of a product, an improvement in the workplace, the reduction or elimination of a workplace hazard, etc. The originator of the ECO coordinates with each function affected by the change to assure that the ECO contains the correct information. The engineer is responsible for assuring the accuracy of the information added to the data base.

1 Methods Improvement

A. BRIEF REVIEW OF A METHOD AND THE METHODS OF IMPROVEMENT

A method is "a procedure, technique, or way of doing something, especially in accordance with a definite plan" [16] whereas methods or process improvement is a continuous process of examining each process with a goal of continual improvement. There are various methodologies which can be used including lean manufacturing, Six Sigma, Value stream mapping, and Kanban [17].

B. EXISTENCE OF SITUATIONS THAT DEMAND IMPROVEMENTS WITH RECOMMENDED SOLUTIONS

1. The company is not profitable or is not as profitable as other companies in the same industry.

Reasons for an unprofitable or less than desired profitable company include a gross income that is insufficient to cover fixed costs as rent, interest expense, etc. Gross income that cannot pay fixed expenses may result from sales prices that are too low and manufacturing costs that are too high.

Sales prices need to be compared with those of competition and adjustments made if possible.

Excessive manufacturing costs may result from noncompetitive prices and terms, the use of material with specifications that exceed those required, the use of inferior materials that can result in excessive rejects and rework cost, excessive direct labor costs due to lack of proper supervision, trained and properly motivated employees, the lack of production labor standards that will determine standard direct labor costs and an evaluative tool to detect opportunities for improvement, the lack application of lean manufacturing technologies, excessive costs of quality due to a lack of quality standards, excessive packaging costs, excessive costs of workplace injuries and accidents, excessive inventory levels of raw materials and work-in-process, and the failure to evaluate all overhead costs for opportunities for improvement.

Each of these possible reasons must be evaluated to determine one or more potential solutions to include a timetable of implementation, the cost of implementing each possible solution, and the amount of cost reductions that will result from each potential solution to enable management to prioritize an implantation schedule based on the needs of the organization.

DOI: 10.1201/9781003153412-1

2. Repeated late or early deliveries

The reasons for late or early delivery under control of the company include an incorrect bill of materials, inaccurate inventory levels, inaccurate information used for scheduling, a delayed receipt of a critical raw material, inaccurate production standards which include both a setup time in hours per set and run time in hours per unit, and waste percentage that deviates from the mean. Inaccurate bills of materials may result in the lack of required components or excessive inventory. Inaccurate inventory levels could result in production delays or excessive inventory levels. Inaccurate information used for scheduling could result from the lack of required components or excessive inventory levels. Inaccurate production standards will result in an imbalance in production, inaccurate costs of production, value of inventory, and allocation of overhead.

Both early and late shipments must be eliminated for several reasons since either can result in customer dissatisfaction and the potential loss of customers.

One cause of both is inaccurate direct labor standards that has several critical uses, one of which is the establishment of standard direct labor costs. Inaccurate standard direct labor costs can result in inaccurate costing and pricing structure, incorrect decisions concerning changes in a proposed manufacturing process, a vendor change, and other proposed changes.

An incorrect waste percentage will result if either too many units or an insufficient number of items being produced, an incorrect cost of material due to the waste factor that is added to the quantity required to manufacture the item, and an incorrect order quantity that could cause an excess or shortage of material. The cause of waste must be periodically studied to determine strategies for its reductions and eventual elimination.

Scheduling methodology uses the production and setup time from the production routing to determine to begin manufacturing the item as well as to purchase needed components based in current inventory levels and the amount of time needed to replenish the item from the current vendor. Thus if either the setup or run time schedules less time to produce a product than is actually required, then the result will be a late delivery. If the setup and production times in the routing file exceed the actual time needed to set up and produce the item, the result will be an output available for shipment earlier than required.

The solution involves the selection of a project team that will follow manufacturing orders through the manufacturing process from the beginning of the order until the order is shipped. A variety of orders should be followed to insure that all possibilities of causes will be included. The actual results must be compared to the standard times for each aspect of production to determine the variances and the amount of the variances. The percentage deviation from variance will determine the follow-up action that needs to occur.

There are only two reasons for the deviation: an incorrect standard or the time to produce the order, the setup time is excessive or too short, or the percentage of defects is incorrect. The reason for the deviation will dictate the corrective action to follow.

a. A material waste percentage that deviates from the percent allowed in the bill of material will result in a positive material variance if the waste percentage exceeds the percentage in the bill of material and a negative variance if less than the percentage in the bill of material. A variance will affect the material costs, and the inventory balance. The actual waste percentage needs to continually monitored for improvement opportunities.

b. Productivity levels that deviate from standard. A productivity percentage exceeding 100% will result in a negative direct labor variance and products completed earlier than scheduled. A negative direct labor variance means that the actual costs is less than standard. A productivity level less than 100% results in a negative direct labor variance, products that are completed later than scheduled which may cause late deliveries and actual costs that exceed standard. The result would result in a potential loss on that order.

Productivity studies must be made if the productivity deviates from 100%.

If the production standards allow more time than required, the results will be an imbalance in the production schedule, overvalued work-in-process inventory, excessive standard labor costs, the loss of potential sales and potential due to the higher standard labor costs and incorrect decisions when attempting process or vendor changes. See Chapters 3 and 4 for an explanation of the methods used to determine direct labor standards and the need for accuracy.

3. A higher employee turnover rate than the industry average. This could result from improper supervision, below average wages and benefits, the perceived lack of advancement opportunities, and poor working conditions. As Richard Branson explained, "There is no magic formula for great company culture. The key is to treat your staff how you want to be treated."

The proposed solution must begin with the human relations department and can include incentives for increased production, reduced scrap, and a reduction in the accident and incident rates. See additional topics for potential training and reasons for types of training that are available.

4. Disorganization in the workplace

This can result in excessive movement of material and personnel to avoid the clutter, excessive defects due to the potential for damage that may occur as a result of the excessing movement. Disorganization in the workplace is an indication of inadequate supervision. One available tool of lean manufacturing that can be employed to address this problem is 5S, which represents Sort, Set in Order, Shine, Standardize, Sustain. The 5S program is a framework that is designed to stress the mindset and tools to continually improve efficiency and value. The method focuses on the reduction and elimination of all waste. See Reference 24 for assistance.

5. Idle employees who should be working performing various tasks or employees who are not fully engaged in their assigned tasks.

This time, referred to by the author as unforced idle time, could be the result of an unbalanced assembly line or poor communication. Unforced idle time can result in reduced productivity, leading to higher direct labor costs, late shipments, and a decrease in employee morale.

If it occurs on an assembly line, the assembly line needs to be studied to determine which tasks can be reallocated to other operators to achieve a more balanced assignment of tasks to minimize unforced idle time.

If it occurs as a result of poor or misguided instructions, then more effort needs to be made by management to insure that all instructions are well understood and there is no opportunity for misunderstanding.

The author while employed in a union plant observed an employee sitting adjacent to his assigned work station and idle. When asked by the author why he was idle, the operator responded that the assistant plant manager gave him one task and his immediate supervisor within minutes provided him with a completely different task. He further stated that he would do nothing until someone gave him only one assignment. The situation was resolved minutes after the author discussed this opportunity for improvement with the plant manager.

6. There are unsafe working conditions that can include unguarded or improperly guarded equipment, the lack of use of PPE, unmarked or obstructed aisles, the use of improper lifting techniques, the improper loading of fork lift trucks, improper storage of LP tanks, the improper storage and charging of batteries, and unkempt floors and aisles.

The presence of one or more of these conditions in an organization indicates a lack of commitment to the health and well-being of its employees. A culture of safety must be inculcated in this organization and supported by all levels of management. More than sufficient resources must be allocated to effect this change as soon as possible to prevent, or at the very minimum reduce, the likelihood of future accidents and injuries to its employees.

The solution is the implementation of an effective safety program if none exists or the strengthening of an existing one.

7. A monthly graph of inventory turn ratios that is not decreasing. This ratio is used to measure the frequency with which a company replaces its inventory over a certain time. The greater the ratio, the faster the company replaces its inventory. A low ratio is an indicator of a slower moving inventory or the maintenance of excessive inventories.

Calculation of the ratio for a particular period can be made by dividing the cost of goods sold during that period by the average of inventory at cost during that period. The cost of goods sold is the sum of the costs of goods manufactured and the cost of inventory at the beginning of that period less the cost of the ending inventory. The average cost of inventory during that period is calculated by summing the cost of inventory at the beginning and the costs at the end of the period and dividing by 2.

The ratio can be used to determine the average length of time required to sell or use the inventory to manufacture and ship the product. It is computed by dividing the ratio into the number of days in a year. The greater the number, the longer the time period needed to sell the inventory. Companies that employ JIT manufacturing tend to have a higher ratio which translates into fewer days to sell the inventory [18].

This ratio represents an average for all the items inventory. A downward sloping graph of the ratio is an indication that overall inventory levels are decreasing and there may be some items in inventory with a much slower ratio. Calculation of ratios for different items may reveal obsolete items that should no longer be in inventory, or items that need to be ordered more frequently depending on delivery time. To determine the turn over frequency for a specific item, calculate the total of the beginning inventory and the quantity purchased during a certain period and divide the usage during that same period. For example, assume that the quantity in the beginning inventory was 25, the purchased during a certain period was 100, the ending inventory was 60, and the quantity consumed during a one-month period was 50. The results yield a turnover of 1.3 months. This time needs to be compared to the length of time required by the vendor to replenish the order. If the vendor can resupply the item is less than 1.3 months, then that suggests that existing inventory levels are excessive and should be reduced. Adjustments in the reorder point may be required. If a change is made, then the inventory turnover rate and inventory level for this item must be monitored to insure that the changes result in the desired result.

Any ratio calculated must be compared with the time needed for the vendor to resupply the item and necessary adjustments made to continually reduce inventory levels.

   8. The occurrence of any incident, accident, or near-miss.

The ultimate goal in the elimination of all incidents, accidents, and near-misses must be investigated to ferret the real cause to determine the correct solution. This investigation should include verification that current regulations and laws are being followed.

   Methodologies for near-miss, accident, and injury reduction are discussed in detail in Chapter 13. The measurements that need to be taken to validate reductions are discussed in Chapter 14.

   9. The receipt of an OSHA fine or violation

The goal of the organization is to provide a hazard-free work environment for its employees. A violation or fine must be thoroughly investigated to determine its cause. The cause must then be addressed to not only prevent a reoccurrence but also to continue to improve the work environment for all the employees.

10. The presence of any of the deadly wastes.

These wastes are:

   a. Transport – defined as the excessive movement of raw material, work-in-process, or the final product.

      Improvement strategies include a redesign of the workplace and not replacing work-in-process inventories.
   b. Inventory – defined as a quantity of raw material, raw materials, or finished product that is greater than immediate needs.

      Improvement strategies include the acquisition of raw materials only as needed and the elimination of buffers in between processes.
   c. Motion – defined as the movement of people or materials that do not add value to the final product.

      Improvement strategies include the redesign of work stations and the employment of equipment designed for greater efficiency.
   d. Waiting – defined as the time lost due to the inability to move to the next phase in order processing or production.

      Improvement strategies include the elimination of buffers in between manufacturing processes and an evaluation of the actual time consumed from order entry to shipment, in order to determine the times and the actions during the time that add no value to the product. The goal is to eliminate those activities that do not contribute value to the product.
   e. Overproduction – defined as the production of a product or component before it is needed resulting in excessive inventory.

      Improvement strategies include setup reduction strategies and a balanced production line.
   f. Overprocessing – defined as the processes that add value to a product that exceeds customer requirements.

      Improvements include simplifying manufacturing processes and insuring that customer requirements do not exceed the manufacturing specifications.
   g. Defects – defined as the products or components that do not meet customer requirements and must undergo additional processes to be acceptable or discarded.

      Improvement strategies include a redesign of the production processes that produce fewer products that do meet requirements and the identification of defects by cause to prioritize resources that will achieve the greatest reductions in the number of defects [19].
   h. A continual increase in material cost that is not justified.

The author has witnessed several such situations which resulted from insufficient monitoring of the involved functions.

One involved a paperboard mill that produced fiberboard from recycled materials.

The author had a job with a paper converter with a sister company that was a paper mill that manufactured paper from recycled material. The mill received newsprint in railcars but also used cartons that were brought in daily from local suppliers, usually in individual pickup trucks. The process was that the pickup truck would enter the mill's yard, weigh in on scales, drive around to the rear of the mill, unload the corrugated waste, return to the scales for a reweighing, and receive in cash the value of the waste based on the current value of corrugated scrap.

One day the author was walking with the chairman of the mill and told him that the author was going to retire and that the chairman was going to fund his retirement. The chairman looked confused and then the author explained how the chairman would fund his retirement legally.

First, the author would purchase two identical trucks and park them a block away from the mill. He would then fill one truck with corrugated waste, spray water on the waste to increase the weight since the price received was a function of the weight, drive to the mill, get weighed, drive around back but instead of unloading continue driving until he was next to the empty truck, swap tags or license plates and drive across the scales with the empty truck, get weighed, and receive cash for essentially nothing except his time. He would continue this process until he had enough cash to deposit that day to avoid suspicion.

The chairman looked at the author in disbelief. Within several weeks the chairman and the president of the company for which the author worked came into the author's office and presented an opportunity for him. The actual cost to produce paper board had been increasing over the last six months although there had not been a cost increase in any of the ingredients, particularly corrugated scrap which was over half of the cost of manufactured board. The chairman had started thinking seriously about someone using my retirement idea and was curious as to whether if it was being employed.

The author's boss asked him to investigate the situation and report the results. The author would randomly check trucks about a quarter mile from the weigh station and discovered that the person in the weigh station was skimming off the top. The receipt was correct but the amount of change returned was less than correct and since most people do not check their receipt against their change, the difference was not noticed. If the difference was noticed, the person in the booth just explained that they were sorry for the error, apologized, and provided the difference.

Another similar situation occurred in a furniture plant for which the author worked as consultant.

Every Monday during the period that the author worked the plant manager would enter orders for several items that were not loaded onto trailers for shipment. One Sunday the author, as he drove to the plant, noticed a large rental van being loaded with items from the warehouse. An investigation revealed that the plant manager and a maintenance person were "relocating" these items to be sold by them at flea markets at prices that were considerably less than the selling price. These items that were relocated on Sundays were the items that were not loaded onto the trailers to be shipped to actual customers. This practice of relocating furniture by these two employees had been occurring for at least several years. The result was jail for the employees and the failure of the company resulting in a loss of over 200 jobs.

To quote a favorite proverb of Vladimir Lenin, *doveryai, no proveryai*, which translates as "trust, but verify," made famous by President Ronald Reagan. This phrase was repeated many times by President Reagan during various meetings with the Russian leader Mikhail Gorbachev [20].

C. REASONS FOR THE EXISTENCE OF AND THE NEED FOR CONTINUAL IMPROVEMENTS

The need for method improvements results from a variety of reasons including decreasing profitability, an increased application of technology, a change in customer expectations and requirements, and a change in regulations.

Several examples will be presented in detail to illustrate methods improvements that will include the results obtained. Proposals for improvement will require different studies and levels of measurement. Some improvements require little formal measurement and can be implemented easily.

D. CRITICAL QUESTIONS TO ASK BEFORE ATTEMPTING METHODS IMPROVEMENTS

1. What is the current method?

It is one that was established when the operation was initially organized. Over time there may have been a few changes as a result of management changes but most managements operate under the philosophy "if it is not broken, then leave it alone." The problem with this is one does not know how much additional productivity can be achieved by an evaluation of each process to determine its need.

Examine the processes before and after the current one to determine if one or more processes can be combined that will reduce the total direct labor costs. Would cellular manufacturing be applicable for this process?

2. Is improvement in the current process possible?

Improvement is possible in every method and process. Reasons for process improvement vary and include the reduction of the costs of manufacturing, overhead, and transportation; an improvement in the working environment; the elimination of hazards in the workplace; a reduction of raw material and work-in-process inventory; a decrease in employee turnover; an improvement in the quality of goods produced; an increase in competitiveness in the market place and increased sales; better compliance with various regulations; and increase in profitability.

The use of various types of measurement will insure that these method and process improvements be used to measure the effectiveness and consistency of improvements.

3. What are some of the tools that can be utilized to effect improvement?

Each of these tools will be discussed in detail later in the chapter. These tools include value stream mapping, Pareto analysis, control charts, statistical process control, fishbone diagrams, brainstorming, scatter diagrams, and process mapping.

4. Why must there be improvement?

Improvement must occur if the organization is to continue to exist. This is true of any organization, whether for profit or nonprofit. Improvement can take many forms, such as in sales growth as measured in dollars and tonnage, number of employees, the number of shifts it operates, the number of shipments, the growth of the rate return, and numerous other financial measurements. The important thing is that the organization must continue to grow in order to survive.

5. Who can initiate a study to investigate a possible improvement?

Anyone can initiate or suggest a study. There must be an existing procedure to insure that the suggested improvement receives an appropriate evaluation and study. The person who initiated the improvement must be kept apprised of the progress of the suggestion as the suggestion progresses through its stages.

This procedure is the engineering change order process [21].

E.   EXAMPLES OF SUCCESSFUL AND UNSUCCESSFUL IMPROVEMENTS

1. Examples of successful improvements
a. Figure 1.1 represents a sanding department that contains ten identical work stations that manually sands chair frames after assembly. The chairs are manually transported from the assembly work station to the elevator which transports them to the sanding department. A material handler removes the frames to be sanded to an available operator. The sanding department consists of a maximum of ten operators who manually lifts the chair two feet from floor level onto the work station. The chair is sanded using a pneumatic sander. The sanding process requires that the chair be manually repositioned often so that all edges and surfaces can be properly sanded. The process uses several grits of sandpaper to insure that the chair is appropriately sanded for the finishing operation. An operator manually moves the chair from his workstation into the finishing department to the quality inspector located at work station 10. Each of the workstations in the hand sanding department is identical.

One day the author noticed, while conducting a study in another adjacent department, one operator in the hand sanding department, currently assigned to booth 2, was wearing a brace on her ankle due to a recent accident. This operator was manually moving a completed chair across a concrete floor approximately 100 feet to the

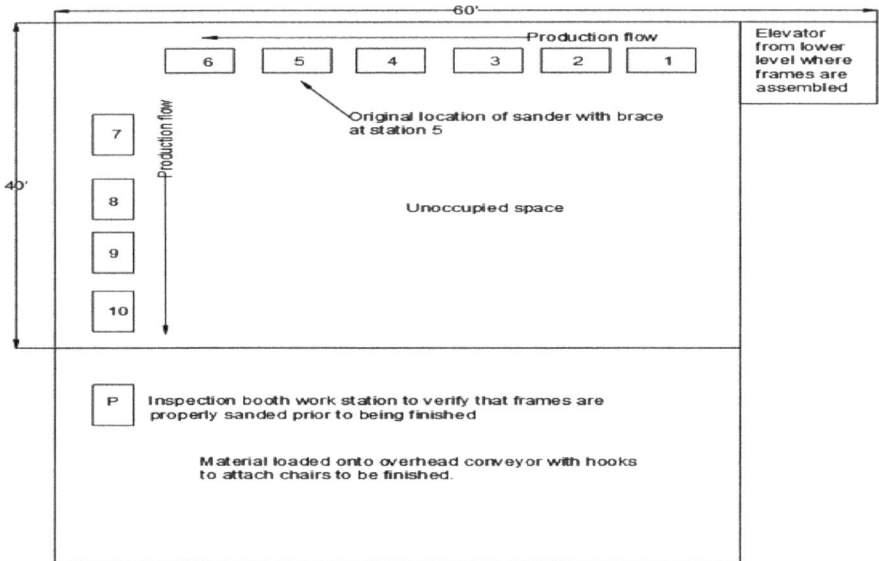

FIGURE 1.1 A hand sanding department with identical work stations with the location of an operator with a brace on one leg.

quality inspection workstation with a brace on one foot. The author then realized that workstation 2 was not currently in use. The author then suggested to the supervisor that the operator be relocated to the available booth to reduce the number of steps required after the task was completed.

This situation provides an excellent opportunity to realize opportunities for improvement. This situation exhibited waste in several areas:

1. The transportation of frames to be sanded and after sanding
2. The elevation of frames off the floor onto the workstation
3. The subsequent lowering of the sanded frames onto the floor
4. The unoccupied space that existed in the sanding department. Sequential operations should be located as close together as possible.

The evaluation of the process to determine the PPE needed. Possibilities include hearing protection, gloves with extra cushioning, and a mat designed for employees who perform a majority of their tasks while standing. Ergonomic studies need to be made to determine what tasks could as easily be performed sitting.

b. The results of relocating the hand sand operator from booth 5 to booth 10 can be seen in Figure 1.2.

Possible improvements include the relocation of the workstations in the sanding department to reduce the transportation time and the readjustment of the heights of

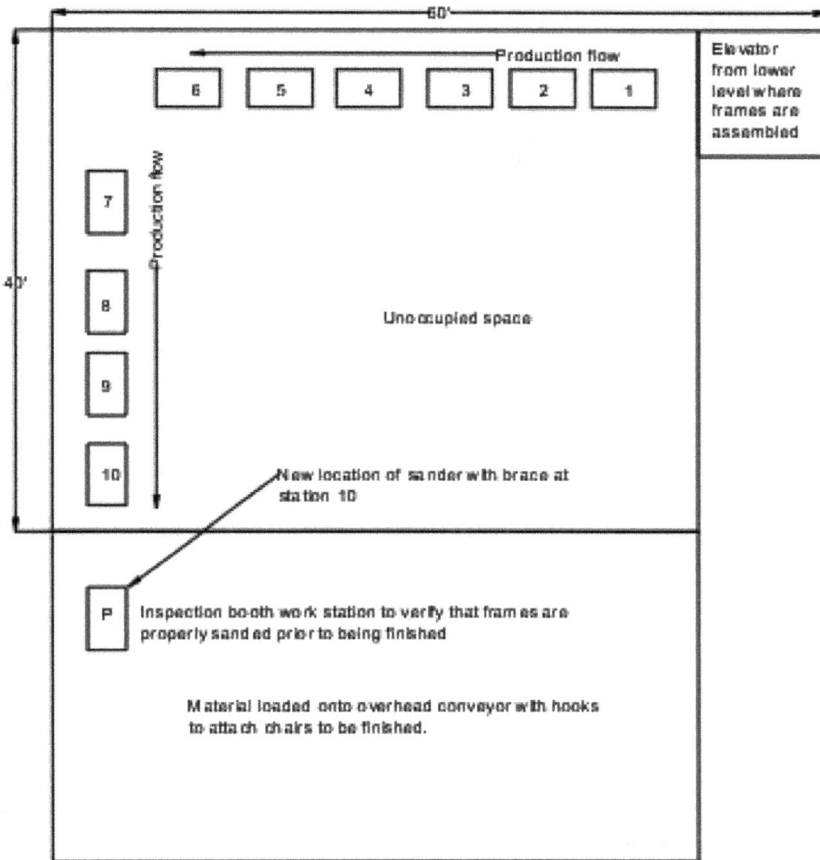

FIGURE 1.2 Original location of operator with brace before relocation to station ten to reduce the amount of travel due to the brace.

the workstations to eliminate the vertical movement of frames. Each potential solution will require different 5 acquired by various methods to insure that the savings that result exceed cost of implementing the changes needed to achieve the desired results.

Primary reason for success here is that the author used a process map to illustrate to management the excessive motions that each operator had to undertake to complete their assigned task. These excessive motions resulted in unproductive tasks and increased the risk of injury from lifting and tripping due to an uneven floor.

    c. An example of another successful improvement and the reasons for its success due to process redesign – lengthening of hoses used for fueling vehicles (Figure 1.3)

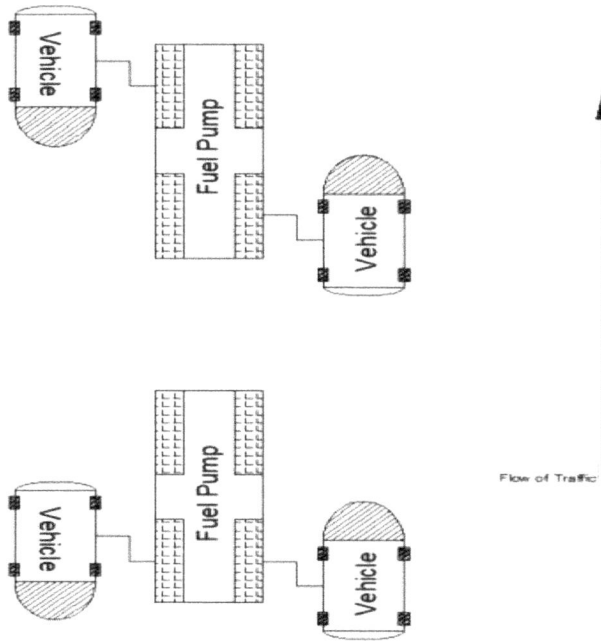

FIGURE 1.3   Queued vehicles awaiting available open pump before lengthening of hoses.

As can be seen in Figure 1.4, the lengthening of gasoline hoses that dispense fuel to enable a vehicle to be fueled from either side of the pump increases the utilization of the pump and decreases the wait of a potential customer.

The situation prior to lengthening the gas hoses can be seen in Figure 1.3.

As Figure 1.3 illustrates, an increased utilization of servers due to an improvement as a result of the application of queueing or waiting line theory.

Lengthening the dispensing hose converted each pump from a single server model to a multiple server model since a vehicle is able to be refueled from either side of the pump. Advantages of a multiple server model over a single server model are a reduction in the average number of customers waiting and a reduction in the average time a customer needs to consume to fuel his vehicle [22].

The reasons for success of this operation were that management used work sampling to determine the percentage of time a pump was idle due to a driver waiting on account of unavailable space on the correct side of the pump and the percentage of time that a pump was idle due to the absence of customers. A comparison between the two percentages resulted in the management's decision to lengthen the hoses to enable a vehicle to be fueled from either side of the pump to improve pump utilization.

Another example of an improvement and the reasons for its success due to process redesign – the process of oil and filter changes at an automobile dealership.

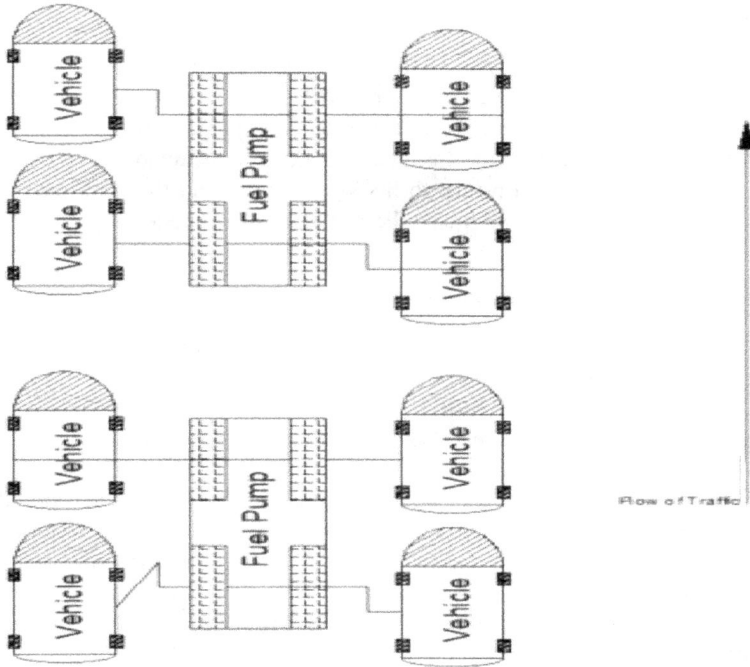

FIGURE 1.4 Queued vehicles after lengthening of hoses to reflect that vehicles can be fueled on either side of the pump reducing wait time and increasing throughput.

THE CURRENT PROCESS

1. The technician would locate a car to be serviced by looking on the dashboard of cars parked on the side of the dealership. The cars to be serviced were parked in an available space.
2. After locating and driving the car to be serviced into the shop and onto the lift, the technician traveled approximately 50 feet to the parts department to request the number of quarts of oil needed and an oil filter.
3. The parts technician, who may or may not be immediately available, would research the part number for the filter, retrieve it from the appropriate shelf, walk up a flight of stairs, select a brand of oil from the four that were available, return to the parts desk, and hand the material to the technician.
4. The technician would perform the needed service on the car, complete the work order stating that the work had been done and sign it, lower the car, submit the signed work order to the office, return the car to an available space, locate another car awaiting service, and repeat the process.

ANALYSIS OF PROCESS

The author performed an analysis of the usage of filters and discovered that over 80% of the filter changes involved only three different filters. The number of different oil filters in inventory exceeded 20. The results of a simple time study taken for further analysis revealed the time that the technician consumed to go to the parts room, wait for the parts, and return to his workstation, and the time for the parts technician to service the car and locate the next car in the queue.

MODIFICATIONS THAT RESULTED FROM THE ANALYSIS

1. Flow-through racks containing two brands of motor oil with the same viscosity were installed in front of each oil changing station.
2. Racks containing the most used oil filters were placed in front of each oil changing station. Signs were placed in front of each filter indicating the brand and year that required that particular filter.
3. Kanban cards were also placed in front of each type of filter and brand of oil. As the technician removed a filter and oil, he compared the remaining quantity of each item with the remaining in the racks and if either quantity equaled or was less than the reorder point on the green Kanban card, the technician took the card to the parts technician or to the parts department clerk who entered an order for the needed parts.

THE PROCESS AFTER METHODS IMPROVEMENT

1. The technician locates a car to be serviced by looking on the dashboard of cars parked on the side of the dealership.
2. The technician, after driving the car to be serviced into the shop onto the lift, obtains the correct filter from the rack and the number of quarts of oil needed, and services the car.
3. The technician completes the work order verifying that the work had been done, signs it, and submits the signed work order to the office; then he returns the car to an available space and locates another car awaiting service.

PROCESS STEPS ELIMINATED

1. The automobile technician traveling to the parts department and returning.
2. The parts technician writing down each individual order from the automobile technician and retrieving the parts from the needed locations.

OPPORTUNITIES FOR FURTHER IMPROVEMENT

1. The assignment of completing repair orders for oil and filter changes to one service order technician to improve this process. Currently there were four service order technicians who could complete work orders resulting in the

likelihood that there were more work orders for oil and filter changes than could be completed.

2. A scheduling system to inform each automobile technician the location of the next automobile to be serviced and a service sequence based on first in, first out.
3. A system to evaluate the productivity of each automobile technician.

The reasons for the success of this improvement were that the author developed a process map and took a simple time study to document the time spent performing each of these activities. The author made a presentation to management to demonstrate the amount of time consumed in performing the unnecessary tasks and the excessive inventory.

    c. Another example of a successful improvement

THE CURRENT PROCESS

1. Clients would line up at clerk's desk for service.
2. If client needed a shot, the clerk would walk client to end of hall while clients in line waited.
3. If client did not need a shot, the clerk completed form for service needed for the next client, who would wait in hall on bench if needed.
4. The clerk serviced the next client (Figure 1.5).

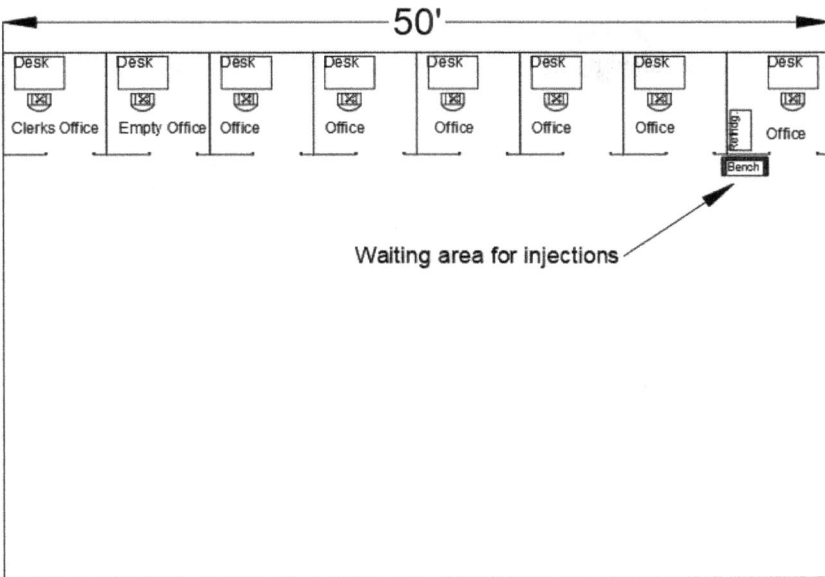

FIGURE 1.5 Layout of health department clinic that to demonstrate the distance that the clerk must travel to accompany patients who need vaccinations.

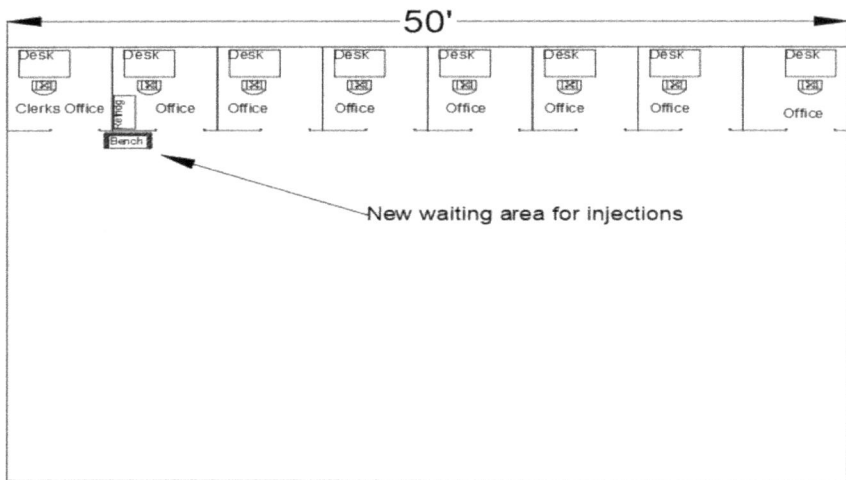

FIGURE 1.6 Layout of health department after the relocation of the office that administered vaccines to demonstrate the amount of walking eliminated resulting in the saving of time and increased patient throughput.

ANALYSIS

A study conducted at a public health department to improve productivity revealed that bottlenecks occurred during the registration process. If a client needed a shot, the clerk would escort him to the waiting room at the end of the hall and would tell him to wait until called by the nurse. The clerk would then return to the front desk and serve the next client in line.

The reason provided for escorting clients to the end of the hall was that the household refrigerator that contained vaccines and other medications was in that office. The suggestion was made to relocate the refrigerator. The refrigerator was unplugged, rolled into the empty office adjacent to the clerk, and plugged into an available outlet for immediate use (Figure 1.6).

Modifications included the relocation of the refrigerator into the adjacent office and the waiting bench into the aisle outside the previously vacant office.

THE PROCESS AFTER METHODS IMPROVEMENTS

1. Clients would line up at clerk's desk for service.
2. If client needed a shot, the clerk would notify the nurse in the adjacent office.
3. If client did not need a shot, the clerk completed form for service needed for the next client, who would wait in hall on bench if needed.
4. The clerk serviced the next client.

Process Steps Eliminated

The clerk escorting the client, who needed a shot, to the end of the hall and returning. This resulted in increased utilization of the clerk. The percent increase could not be measured due to the variability in the flow of incoming clients.

Opportunities for Future Improvements

Reduce variation in the flow of incoming clients by scheduling appointments.

The reason that this improvement was successful was a flowchart that the author used to demonstrate the amount of time lost due to the excessive motion of the clerk. Another reason resulted from the lack of critical thinking earlier – no one asked why this current method existed and if there was a better method.

   d. Another example of a successful improvement

The process began with the raw material, which is tubular steel of a designated strength and a specified diameter and length. The final item in the process was a metal chair frame that was paint ready. The final product was not loaded onto a paint line at that time because the powder coating operation was performed in batches by color due to the lengthy and costly changeovers. One operator began with the raw material and performed all the necessary tasks sequentially. The machines needed to accomplish each task were arranged in the form of a semi-circle. The operator would begin the next task only upon completion of the current one.

The manufacturing process before the cell was created consisted of the initial operator being supplied with a tote bin of tubular steel by means of a fork lift operator; an operator removing a length of tubular steel from a tote bin, performing the designated task, placing the completed component into a different tote bin, waiting for the material handler to remove this tote bin and relocate it into position for the next operation. The material handler would scan the tote bin with the number of items contained, then after the tote bin was relocated to enable the next operation to occur, rescan the tote bin after the raw material had undergone a process, the part number changed to a different one with a bill of material which was the raw material, that is, the tubular steel of a specified length and thickness, and a routing with two steps. The first was retrieving the raw material in the tote bin from the warehouse and taking the tote bin to a specified machine in the machine department, and the second step was performing the designated task and placing the component into a different tote bin.

This particular item consisted of seven different manufacturing processes, each operation requiring two tote bins, one containing components to be processed and another for components that had completed processing. Each task during the manufacturing process required the generation of a work order with a bill of materials and routing. After completion of each task the manufacturing order was scanned into the computer to update the master schedule.

```
                         ┌──────────┐      ┌──────────┐
                         │Operation │      │Operation │
                         │4         │      │5         │
                         └──────────┘      └──────────┘
                                                         ┌──────────┐
              ┌──────────┐                               │Finished  │
              │Operation │                               │Product Bin│
              │3         │                               └──────────┘
              └──────────┘

     ┌──────────┐
     │Operation │        Order of operation in MFG Cell.
     │2         │        Operation 1, Bend straight piece of tubular steel into a U shape.
     └──────────┘        Operation 2, Bend each side of the U shape into an arm shape.
                         Operation 3, Punch holes into the bottom of each arm shape.
  ┌──────────┐           Operation 4, Punch holes into the bottom of each U shape.
  │Operation 1│          Operation 5, Punch holes in side of each U shape to attach to frames
  └──────────┘


  ┌──────────┐
  │Raw       │
  │Material  │
  │Bin       │
  └──────────┘
```

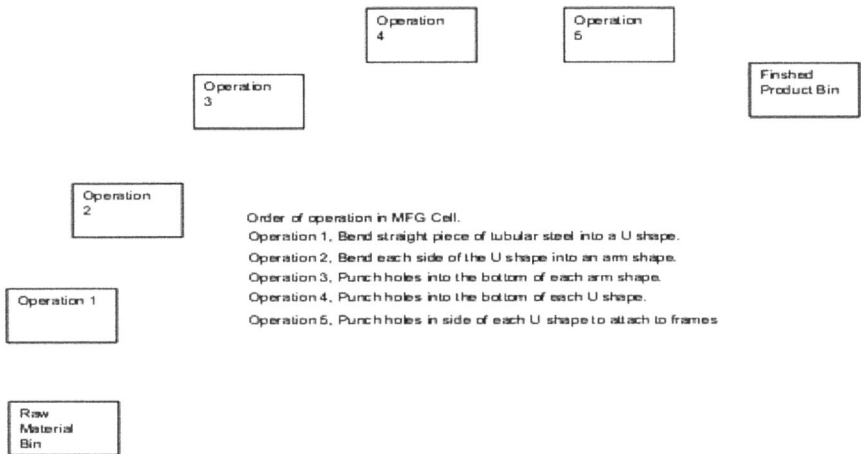

FIGURE 1.7 Layout of a manufacturing cell that resulted in reducing direct labor costs, work-in-process inventory, floor space needed, and an increase in throughput.

After establishing the cell, only one manufacturing order was generated with a bill of material and routing. The routing consisted of two tasks, the retrieval of the raw material from inventory and performing the required operations. There was only one work order to be scanned. The number of tote bins required during manufacturing was reduced from seven to two, one containing the raw material and the other the completed component that needed additional processing. The floor space required was reduced by 40% due to the elimination of tote bins. Work-in-process was reduced drastically since all the intermediate processes were eliminated. Manufacturing throughput was increased (Figure 1.7).

The reason for this successful improvement was that the author compared the standard hours using the existing method with the standard hours using cellular manufacturing. This was possible due to the elimination of the aside time into a tote bin for each operation since, after each process, the completed part was placed into the adjacent machine for processing. A flow diagram of the new process drawn to scale demonstrated that the footprint of the metal processing department would be reduced by 40%.

e. Another example of a successful improvement

The engineers in one furniture company in which the author worked joked that the furniture was worn out by the time it was received by the customer. For example, the cutting and sewing department was located in a building approximately ¼ mile from the upholstery department. The reason was that there was insufficient parking near the upholstery department. A 16-foot van with a full-time driver was employed to transport sewn covers from the building where the covers were sewn to the upholstery department where the covers were used. Exacerbating the situation, the sewing department was on an individual incentive system, necessitating that every cover counted be credited to the sewer, and placed into a laundry basket before being loaded onto the van for delivery. After the van was received, the covers were removed from the laundry basket, again counted and distributed to the next available sewer.

A study was made of the cost of the full-time driver and the truck, which included the cost of both loading at the sewing department and unloading at the upholstery department. A contractor was then contacted to quote a cost for providing a sufficient number of parking spaces for the cutters and sewers. Payback was less than two years and the parking spaces were added as soon as the contractor could.

The reason that this improvement was initiated was that the van transporting the cut covers to the sewing department was involved in a minor traffic accident, resulting in a delayed delivery to the sewing department. The delay proved long enough for numerous sewers to incur unforced idle time, which resulted in production interruptions in subsequent processes. Overtime was required to compensate for these interruptions. The accident and the resulting costly downtime caused management to examine and study the reasons for the covers to be transported to a distant department. To quote Da Vinci, "we look but do not see" [23]. The reason is that our brains are not engaged as an activity is observed, which results in no memory of having observed that activity at a later date.

f. Another example of a successful improvement

One furniture company for which the author worked received foam and a crew of three stored it in the warehouse in racks that could be as tall as eight feet. Foam is the most voluminous part of furniture but is relatively inexpensive. Then a different crew of three would retrieve the foam and take it to the appropriate department, seats or backs. Due to the height of the racks, a ladder was required and a person was needed to hold the ladder due to safety requirements and the bulky nature of the foam being retrieved.

The person who retrieved the foam would first scan the foam indicating the material transfer. They would then transport the material to the department, locate the supervisor to determine which workstation or workstations should receive the foam, and then return to the warehouse.

The reason that this improvement was successful was the presentation of a flow diagram and a manual simulation of employee movements during this process.

This improvement resulted in the reduction of indirect labor costs, an increase of available storage space, and the reduction of an accident or injury due to elimination of the use of a ladder.

g. Another successful example

This example illustrates the importance of the application of critical thinking and that a bottleneck operation can exist in any process.

The author had the opportunity to consult with a company that manufactured prefabricated steel buildings. These buildings could vary from a simple shopping center to an airport hangar to a multilevel church and everything in between. The buildings were shipped to the site of erection, FOB plant, throughout the United States and sold to contractors who would then erect the buildings on site. Contractors were required to add the interior components as insulation, heating, air conditioning, and the roof since that information was not provided on the initial documentation.

The process of providing consisted of having a structural engineer determine the necessary components, a mechanical engineer draw the building using an automated drawing system which would determine all the components needed to construct the building, a quality engineer certify the drawing, a costing engineer determine the delivered cost of the building, and the president review the documentation and forward a quote to the contractor.

Depending on the desire of the contractor at the time of the quote, the contractor may discount the quote as much as 25% to the final customer in order to obtain the contract. Less than 40% of the quotes given actually became a contract. The length of time between the issuance of a quote and the receipt of a contract varied from a minimum of 4 weeks to a maximum of 12 weeks depending on the complexity of the building.

The president called and explained the situation and asked for help. He stated that his engineering staff was overloaded and he was forced to pay overtime, shipments were late and made with items missing due to the overworked resources; the company was facing potential losses for the first time in history and was seeking assistance.

He further mentioned that the company maintained many years of data concerning various information about the building for which they had provided contracts. This information included basic information as length, width, height, the number of parapets (an upper extension of the wall), the zip code, the intended end use of the building, the proximity to the nearest fire station, etc. This data was numeric as well as alphanumeric.

The author was in graduate school during this time working toward a PhD in industrial and systems engineering and thought that multivariate analysis could be used to address this situation. Multivariate analysis is one of many statistical techniques available, but the author felt that it was most appropriate since it focused on the interrelationships among several variables.

The goal of the project was to use the data to develop a tool that could be used with at least a 90% degree of accuracy that could predict the cost of a building with as little information as possible. The president realized that the constraint was due to the existing process of determining quotes and was seeking a method to expedite that process. The data in multivariate analysis may be numerical, categorical, or a mixture of the two. The data was not time sensitive since no variables included any cost-sensitive data as money. For example, production time was expressed in man-hours per square foot of shipped building. All the variables were able to be decomposed into man-hours per square foot of shipped building.

Data that was available spanned the most recent ten-year period. Turnover during this period was essentially nonexistent, which increased significantly the confidence level of the data entry into the existing excel spreadsheet. Using the first eight years as a test, the author entered the data into the statistical program used at his institution. The results indicated a high correlation among height, width, length, and zip codes. Further analysis and elimination of variables as a result of low correlation resulted in the ability to predict the building costs of a prefabricated building with a 95% degree of accuracy using only four variables – length, width, height, and the zip code. The zip code was needed to predict weather issues as the location of the proposed building in an area subject to tornadoes or flooding.

The last two years of data were used to verify the model, which was proven to be 95% accurate.

This prediction tool enabled this company to eliminate the immediate constraint or bottleneck of determining materials needed for providing quotes to contractors. This resulted in the elimination of late shipments, overworked resources, and items missing from shipments. The use of this tool enabled the company to devote resources to insure that shipments were shipped on time and contained all the needed components with simple instructions. The procedure in place now is that when a request for a quote for a new building is received, the information is entered into the prediction tool and then the existing database is checked to corroborate the estimate before the quote is provided to the customer, usually the following day.

The company was able to significantly increase the turnaround time for quotes from a mean of over four weeks to one day. The result was increased sales, a dramatic decrease in complaints due to missing or incorrect components and an increase in customer satisfaction. All of this was achieved because someone recognized there was a better way and devoted the resources to locate and implement it. As Thomas Edison said, "There is a better way – find it" [24].

2. Unsuccessful examples

a. The reduction of finished goods inventory of printed labels

A company which manufactured and printed labels on narrow fabric maintained various quantities of finished goods inventory at the request of the vendor. The amount in finished goods varied from several hundred thousand to a million. The labels varied in width from ¼ inch to 1½ and in length from ¾ inch to 3½ inches in length. The various processes included weaving the yarn into the widths needed, adding starch to stiffen the woven fabric to enable it to be properly printed, folded and machine packed in cartons of one thousand.

Orders were received and paid for in units of thousands (K) but the company required that the software supplier convert the software accept units since the management believed that would be too confusing for the employees if the employee saw an order for 14K rather than 14,000, resulting in numerous errors.

CURRENT PROCESS

1. Customer enters order to customer service.
2. Customer service place order with plant.
3. Plant prints, folds, and packages labels.
4. Plant warehouses labels in mezzanine above manufacturing floor.
5. Customer service enters makeup orders to plant for any discrepancy between initial order quantity and quantity produced.
6. Plant prints, folds, and packages labels from makeup order.

7. Plant warehouses labels in mezzanine above manufacturing floor.
8. Plant ships order.

ANALYSIS

The company maintained a minimum level of finished stock for certain items at the request of major customers. These labels were stored on the second floor of the mezzanine above the manufacturing floor. Because there was no coordination between orders that were being produced and those being shipped, recently produced items could be stored in inventory at one end of a shelf as that identical item was being removed for shipping from the other end of the same shelf. The items being added to inventory may represent a new customer or replacement order. The company did not have a memo of understanding with any customer stating that the customer would exhaust existing finished goods inventory of a particular label before any changes were made in that label. As a result, if a customer revised a label, the company maintained a finished goods inventory with zero sales value.

A frequency distribution was made for each major customer of order frequency by item. A Pareto analysis was made for each major customer that identified items that which comprised 80% of order quantity by item. A study was then conducted to determine the mean production time and standard deviations as a function of quantity. The minimum and maximum production time for each item and the quantity produced were also determined.

This comparison indicated that all production time for any item for all production quantities was less than the allowed delivery time. These results were presented to management with these prioritized objectives:

1. Reduce finished goods inventory with the ultimate goal of its elimination. This goal was supported by the analysis indicating that production time was less than delivery time and the absence of memos of understanding.
2. Eliminate makeup orders. This objective was suboptimal and supported the same analysis, but the implementation would be an improvement resulting in a decrease in the level of finished goods inventory.
3. Educate everyone as to the need for continuous improvement essential to remain competitive and the types of information needed, the source of that information, and the steps in the decision-making process.

Management agreed that continual improvement must occur. The results of the meeting resulted in the elimination of the process of issuing makeup orders and then evaluation of the impacts of this decision on inventory levels and the percentage of on-time and early deliveries. A follow-up management meeting was to be scheduled in three months.

PROCESS AFTER CHANGE

1. Customer enters order to customer service.
2. Customer service place order with plant.

3. Plant prints, folds, and packages labels.
4. Plant warehouses labels.
5. Plant ships orders upon receipt of an order.

WHAT POTENTIAL CHANGES SHOULD BE INVESTIGATED FOR FURTHER IMPROVEMENTS?

1. Enabling primary customers access to inventory quantities to enable customers to order needed items electronically to reduce errors and the need for customer service personnel involved.
2. Reducing setup times in printing presses to increase throughput, reduce downtime with the ultimate goal to eliminate finished goods inventory.
3. Investigating work-in-process inventory with the ultimate goal of reducing it to minimal items and minimal quantities.

Unfortunately, the results of the follow-up meeting are unknown since the engineer who initiated the studies left for another position. The company ceased operations in that facility soon afterward resulting in the loss of over 400 jobs and the spiraling effects of those job losses. This loss may or may not have been prevented but it underscores the need for continuous improvement.

b. The relocation of natural gas meters from under to the outside of a house

TITLE		Natural Gas Meter (plan view)				
Figure Number		1.8	Drawn By	JD	Date	9/28/2020
Natural Gas Meter						
Checked by	JW	Date	2/21/2021	ECO #	978-022	

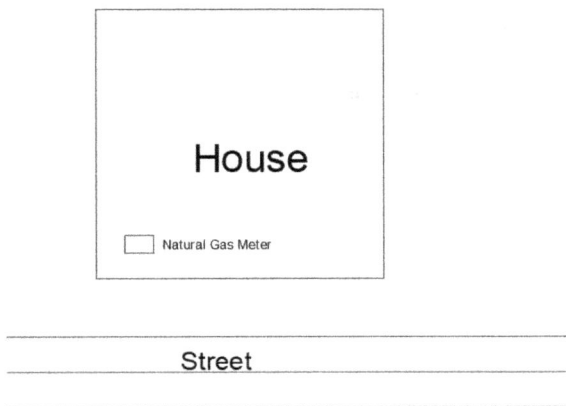

House

Natural Gas Meter

Street

FIGURE 1.8 Layout of house with the natural gas meter under the house

TITLE	Natural Gas Meter (plan view)				
Figure Number	1.9	Drawn By	JD	Date	9/28/2020
Natural Gas Meter					

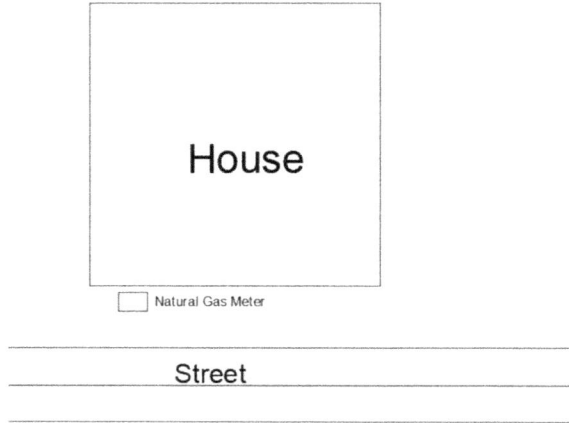

FIGURE 1.9 Natural gas meter after relocation of natural gas meter located in front to facilitate reading of the meter

The current process required that steps be performed in sequentially

CURRENT PROCESS

1. Obtain orders (appx four to six) from the company at the beginning of the workday
2. Proceed to the first address and confirm address and that meter was under the house
3. Crawl under the house to locate the gas meter
4. Disconnect natural gas pipe to house
5. Verify that gas is disconnected
6. Loosen fittings to meter, remove meter from under house
7. Reconnect incoming gas line from street to the meter
8. Crawl under house to obtain measurement from outgoing meter nozzle to where the point in the line under the house where the outgoing line was connected
9. Convey that measurement to the crew to enable them to cut and thread the lengths of cast iron pipe needed to complete the connection
10. Add the necessary components, tighten securely.

11. Turn on the gas and test all joints with a solution of soap and water.
12. Retest all joints with solution of soap and water
13. Clean up job site and relocate to new job site

ANALYSIS

The crew consisted of a plumber, three laborers, and a supervisor who was a licensed plumber. The crew was hired to relocate gas meters from under houses to a position on the side instead of the front to prevent detraction from the hose and to facilitate reading of the meter. The supervisor would receive at least four orders daily for houses in various locations throughout the metropolitan area. Arrival to a new site required an average of thirty-two minutes; the actual relocation required an average of twenty-four minutes,

The author, who was one of laborers, proposed to the engineering manager that productivity would be increased if the crew eliminated the travel between assignments by remaining on the same street until all houses on that street have been serviced. Productivity would be increased at least 100 percent since the average time for meter replacement (24 minutes) was less than the average travel time (32 minutes). The proposed new process would eliminate the sequential performance of tasks to enable several tasks to be performed simultaneously resulting in greater productivity.

PROPOSED NEW PROCESS

1. Proceed to the first address and confirm address and that meter in under the house
2. Crawl under house to locate the gas meter and to verify that gas meter is under the house
3. Disconnect natural gas pipe to house
4. Verify that gas is disconnected
5. Loosen fittings to meter, remove meter from under house
6. Reconnect incoming gas line from street to meter
7. Crawl under house to obtain measurement from outgoing meter nozzle to where the point in the line under the house where the outgoing line was connected
8. Convey measurements of pipe needed to the crew to enable them to cut and thread the lengths needed to complete the connection
9. Add the necessary components, tighten securely.
10. Turn on the gas and test all joints with a solution of soap and water.
11. Retest all joints with solution of soap and water
12. Clean up job site and relocate to new job site

Management did not implement the new process because meter relocation was a project that was only performed during the summer due the possibility of severe winter weather and to prevent any disruption in service. The engineering manager agreed to adopt the new process at the next implementation of this project. Benefits would include at least a 100 percent in productivity and a reduction in the time required to read meters, least a 400 percent increase in productivity. This does not include the increase in speed and accuracy in the reading of meter readings.

This improvement attempt was not successful due to the lack of commitment of management. Management has a process that worked and failed to realize the potential of the proposed improvement.

F. CRITICAL QUESTIONS TO ASK FOLLOWING THE IMPLEMENTED METHODS IMPROVEMENTS WITH SOLUTIONS

1. To what degree were the goals of methods improvements achieved?

Results of improvement cannot be determined unless there is a defined metric or metrics that were established prior to the initiation of improvement efforts, and without such a set of metrics, it is not possible to ascertain the actual improvement achieved or the degree of improvement.

2. What were the main reasons that prevented the attempted methods improvement from being a complete success?

The primary objection that the author faced when attempting methods improvements was a reluctance to change from the existing to the proposed method. This reaction is to be expected because most people would prefer to retain the status quo since change requires a thinking about and performing a task.

3. How to achieve improved results when initiating methods improvements?
   The following steps should be followed:
   a. Insure that the proposed method works and will yield the expected results. This action must occur before attempting to begin methods improvements. Improvements can be made after implementing the new method.
   b. Request input from the employees to be effected to enable them to understand the reason for the improvements and appreciate the benefits that will accrue to themselves and the organization that will result from the improvements [25].

2 Measurement

A. BRIEF OVERVIEW OF THE CRITICALITY OF MEASUREMENTS AND THE NEED FOR ACCURACY

Measurement is essential to change because it provides proof that one has occurred. Change can be either positive or negative. A goal of industrial engineering is continual improvement which results in positive change but continual improvement also must include the improvement of the steps involved in the decision-making process. A byproduct of being human is the commission of mistakes. Continual improvement includes an analysis of a mistake to prevent its reoccurrence.

B. CRITICAL QUESTIONS THAT MUST BE ADDRESSED

1. What is the definition of a measurement?

The word "measurement" is derived from the Greek word "metron", which means "limited proportion". [26].
    Measurement is:
    The assignment of a number to a characteristic of an object or event to enable it to be compared to other events or objects phenomena.

1. A technique that determines properties of an object comparing them to a standard.
2. Measurements require calibrated tools and provide a quantity. A quantity depicts how much or how many there is of something. A simple example of measurement consists of using a ruler to find the length of an object. The object is whatever you are measuring, the property you are trying to determine is the object's length, and the standard to which you are comparing the object's length is the ruler.

Measurement is fundamental to the sciences; to engineering, construction, and other technical fields; and to almost all everyday activities. Frequently it occurs to product quality and to monitor the efficiencies and productivity of its processes [27].

C. POTENTIAL SOURCES OF ERRORS IN THE USE OF MEASUREMENT

Thus measurements consist of a quantity and an attribute or property. Each is equally important to insure the accuracy of that information. Mistakes have occurred due to the assignment of the incorrect attribute. A spacecraft that was

DOI: 10.1201/9781003153412-2

designed to study the atmosphere on Mars was lost by NASA due to failure to 33convert units of measure properly. The contractor provided data in the U.S. imperial system consisting of inches, feet, and pounds but should have converted those units to the comparable units in the metric which was standard practice for space missions. The navigation engineers at another location assumed incorrectly that conversions had occurred. The contractor supplied information using the English System (pound); the team of NASA navigators expected to receive the information using the Metric System (Newtons). The cost of the orbiter spacecraft in 1999 was 125 million USD [1].

Many errors have been reported in the administration of medication. The Institute for Safe Medication practices has issued several publications to educate healthcare professionals concerning potential areas,

   a. These include a list of confused drug names as Advair versus Advicor and Adderal versus Inderal [28]
   b. the list of error-prone abbreviations, international units whose abbreviation is IU can be mistaken for IV,
   c. the recommendation is to use units, the abbreviation of intravenous, and OXY, oxytocin which can be confused with oxycodone or oxyContin.
   d. The recommendation is to use the complete drug name and not to abbreviate [29].

The results of a study conducted between July 1994 and June 1995 of errors in a hospital that offered a higher level of care committed during the prescription phase of medication administration found that over 2,000 (2,103) were thought to have clinical importance. The major factors that contributed to the errors were knowledge of drug therapy and its application (30%), inaccuracy in calculations and decimal points (17.5%), and incorrect drug name, dosage, and abbreviations (13.4%) [30].

D.   THE REASONS TO TAKE MEASUREMENTS

   1. To Monitor a Process
        These measurements must occur over a long period of time to discern trends and variations that occur due to the process. for analysis to determine the presence and extent of problems to enable an understanding of trends and variation. Opportunities for improvement can be evaluated after the problem has been identified.
        The process of taking measurements cannot interfere with the process itself. It must require that each measurement is repeatable, performed on a consistent basis, and be inexpensive.
   2. Investigative measurements are taken for the specific purpose of obtaining additional information concerning existing problems or the suspected causes of these problems and cannot interfere with the process. The conduction of these investigations will result in minimal or no interruption to the operation.

E. WHAT IS DRIVING THE DEMAND FOR THESE MEASUREMENTS?

The demand is driven by the requirement for continuous improvement in all sectors of the organization. Improvement opportunities exist in the utilization of all labor, increased yield of all materials, a reduction in the amount of floor space required as a result of process redesign, the capital invested in the various material inventories and work in process, overhead expenses as utilities, indirect expenses and as the medical and other associated costs of accidents and incidents resulting in a reduction of the cost of workmans' compensation insurance al

F. HOW WILL THESE MEASUREMENTS BE USED?

1. To determine or revise existing direct labor standards

   These updated direct labor standards will provide input into direct labor costs, individual and group incentive systems, scheduling of the manufacturing process, determining the capacity of a single operation, an assembly line, or the entire facility.
2. To determine material costs by the undertaking of material utilization and waste studies.
3. To determine optimum floor space utilization for the storage of raw materials, work in process inventory, and manufacturing processes.
4. To determine the amount of time and resources needed for indirect labor tasks for manufactured products toward delivery, the unloading, inspection, and storage of raw materials, inspection of work in process, and the performance of various maintenance activities.
5. To assist in the decision of installing a manufacturing process change – Chapter 9.
6. To assist in the decision of making.
7. To assist in the making of a vendor change – Chapter 11
8. To assist in the reduction of seven deadly wastes.

Waste is simply defined as something that does not add value to the product or service and exists in each of these areas. Examples of waste include the following:

a. Transport is the movement of materials to a different location. These materials include raw materials, work in process, and finished products.
b. Inventory of raw material, work in process, and final products cost money to maintain and insure. Inventory requires floor space to maintain which also costs money. Inventory also requires movement to different locations, which represents opportunities for damage during movement. Also, inventory can become obsolete due to technology changes or expiration dates.
c. Motions are the movements either by employees or of machines that are unnecessary. Examples include the bending of an employee to lift an item on the floor instead of rearranging the workstation so that the item is at waist level. This unnecessary motion could result in an injury to the employee.

Evaluations of motions made by employees should be an integral part of the duties of the safety team. Excessive travel between workstations is yet another example.

d. One result is excessive inventory which is costly to maintain and is subject to damage and obsolesce, loss of floor space, excess cost due to its maintenance, possible obsolescence, and damage from handling. Overproduction or producing more than is required results in excessive inventory. Overproduction is also producing a surplus quantity in anticipation that the customer will order the product.

e. Defects are the most costly wastes. A defect in manufacturing will incur repair, rework, or replacement. Defects are to be reduced to the maximum extent possible. A defect is something unexpected or a deviation from expected. A proper design can reduce the number of defects that are produced during manufacturing. The method of achieving this is by increasing the usability of the product through usability testing.

f. Overprocessing could result from several factors as the performance of processes not required, the use of heavier equipment than is needed to accomplish the task, or meeting tolerances that are too tight. A good example is the use of an expensive painting process to paint the underside of a lawn mower.

g. Defects are an obvious waste and each defective item results in additional costs for repair and potentially causes late shipments. The ultimate cost could be the loss of a customer.

h. Another waste is the waste of the talents, education, experience, training, and skills of employees, the use of an overtrained employee represents activities that are not performed.

9. To measure the success of improvements and implementation of lean tools. These tools include 5S, Bottleneck Analysis, Continuous Flow, Gemba, Level Scheduling, Just-in-time (JIT), Kaizen, Kanban, Key Performance Indicators (KPI), Muda, Overall Equipment Effectiveness (OEE), and Plan Do Check Act (PDCA), Poka-Yoke (error proofing), Root cause analysis, Single Minute Exchange of Dies (SMED), Six Big Losses, Smart Goals, Standardized Work, Takt Time, Total Productive Maintenance, Value Stream Mapping (VSM), and Visual Factory [32].

10. To Assist in the Development of Productivity Improvement Programs. Productivity improvement programs designed specifically for:

a. Individual operator improvement which can be achieved during methods improvement during the determination of direct labor standards (Chapter 3).

b. Productivity improvement in a cellular manufacturing process. A cellular manufacturing process is similar to a production line with the exception that all the processes in a cell are devoted to one product.

c. Overall plant productivity improvement can be accomplished by locating and eliminating bottleneck operations. In every operation there exists a bottleneck. The elimination of one will only result in the realization of some other constraint that exists elsewhere. Constraints will never be eliminated; eventually, the cost of elimination of the restraint exceeds the gains from its elimination. This is an application of the law of diminishing returns.

Productivity improvement program in activities not related to manufacturing as improved utilization of the receiving department can affect indirect labor costs. Indirect labor costs are often allocated to direct labor costs as an additional cost to determine standard direct labor costs. Any productivity improvement in nonmanufacturing activities may affect standard direct labor costs if indirect labor costs are allocated to direct labor costs.

11. To assist in an increase in operator skills and abilities which will result in productivity improvements

    Employees are the most valuable asset that any company possesses. Organizations recognize that intellectual capital is as valuable as physical and financial assets.

    Training:

    a. is an organized method to improve the skills, knowledge, and competency of the employee so that he will be able to perform his job functions effectively.

    b. is a tool to develop and increase thinking skills and creativity that results in improved decision-making, customer service, and an increase in overall efficiency.

    c. improves employee satisfaction which results in a lower turnover rate.

    d. results in an increase in productivity and competitiveness due to the improved attitude concerning learning [32].

    e. plays an important role in informing employees on the importance of their jobs as well as providing the information needed to perform their jobs.

    f. has also been shown to create greater loyalty and improved attitudes among employees and to provide assistance to these employees in their personal development and achievement [33].

    g. has experienced an increase in the global average expenditure per employee over the most eleven-year period for which data is available. The annual expenditure per employee can be seen in Table 2.1 and Figure 2.1. The annual increase correlates positively with time (R square equals 0.823) [34].

12. To assist in the development of a cost-effective maintenance program.

    Management goals have increased since the end of World War II and the subsequent rebuilding of industry worldwide, which resulted in greater competition in the global marketplace. Downtime became less and less tolerated. The ultimate goal is to control the cost of maintenance, the reduction of downtime, and the collection of relevant information to enable

TABLE 2.1

This Table Illustrates the Growth of Expenditures of Workplace Training over an Eleven-Year Period

Year	Workplace Training per Employee USD
2008	1066
2009	1081
2010	1228
2011	1182
2012	1195
2013	1208
2014	1229
2015	1252
2016	1273
2017	1296
2018	1299

Workplace Training per Employee USD

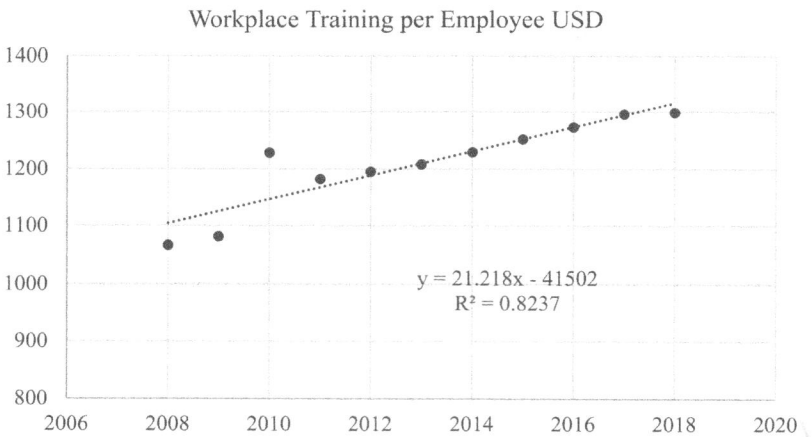

$$y = 21.218x - 41502$$
$$R^2 = 0.8237$$

FIGURE 2.1 This table illustrates the growth of expenditures of workplace training over an 11-year period.

management to make decisions to attain this goal. Maintenance is the proper management of an asset's health and well-being of a physical asset over the expected lifetime of the asset [35].

Maintenance is undertaken to enable equipment and systems to operate efficiently for a minimum of their expected design life. Maintenance is essential due to predicted failure at some time of equipment and systems

There are five types of maintenance programs. These types are represented by the bathroom curve model, reactive, preventive, predictive, and reliability-centered maintenance.

At a minimum, the information needed to implement such a program includes the machine affected, the date, the start and end times of the maintenance, the crew size, the cause of the downtime, a brief description of the work performed, and the items and parts that required replacement [36].

13. To assist in the tools that can monitor progress toward various goals.

    Referred to as development and improvement of key performance indicators (KPIs), they have application in quality; efficiency and productivity, and customer satisfaction. [37]. A set of KPIs has been developed by Agency for Healthcare Research and Quality. These are Prevention Quality Indicators, Inpatient [38].

14. Is this information specified in the process sheets if the company is ISO certified?

    a. Is this information clear and easy to understand?

       It is critical that the information can be understood by those employees who are performing these tasks and for potential employees who in the future may be performing these tasks. The observance of employees as these tasks are being performed and addressing and correcting of understanding and complying with these instructions.

    b. Does the process sheet include the tools, fixtures, quality inspection sheets, and personal protective equipment (PPE) needed for the job?

       The observance of the operator as this task is performed will reveal any discrepancies which must be addressed immediately.

15. What techniques will be used to obtain these measurements?

    Each of these techniques is discussed in detail in Chapter 3.

    a. A formal time study

    b. Work sampling

    c. Getting input from employees

    d. Group timing techniques

    e. Predetermined motioned time systems

3 Integrating Methods Improvements for the Establishment of New Direct Labor Standards or Existing Direct Labor Standards

A. BRIEF INTRODUCTION TO THE NEEDS FOR DIRECT LABOR STANDARDS

Direct labor standards are the basis of many management decisions and have many applications. Some of these applications include the determination of standard labor costs, compute the capacity of an operation, as an assembly line or a manufacturing facility, and to evaluate existing processes for improvement.

Without measurement, there can be no improvement since one must have a starting point against which all future changes can be compared. To be meaningful and useful in various decisions, there must be standardized procedures for establishing these measurements. Industrial Engineering has a methodology to insure that time studies are taken in a consistent manner to enable the practitioner to insure that the desired result was achieved.

B. CRITICAL QUESTIONS TO ASK BEFORE ESTABLISHING NEW DIRECT LABOR STANDARDS WITH SOLUTIONS

1. What is the definition of a standard labor hour for an individual operator?

A direct standard labor hour is the amount of time that a normal operator should take to complete an assigned task given normal working conditions.

A standard labor hour is the product of three variables, the mean time for a particular task in hours, the total of allowances for fatigue, personal time, and delay, and the rating of the operator.

DOI: 10.1201/9781003153412-3

2. What are the primary applications of standard direct labor hours?
   a. To calculate the standard direct labor costs of a product or service.

      The standard labor costs are one of several costs that is used to determine the total costs of materials used to produce or service. Other costs include material and packaging costs and overhead. Transportation costs must be added to those products that will be delivered.
   b. To determine the capacity or a specific period of time, as a shift of eight hours, single process, a production line or the entire organization.
   c. To establish incentive systems which can be based on an individual or a group of operators and are discussed in detail in Chapters 5, 6, 7, and 9.
   d. To serve as a basis for determining other costs.
      1. To allocate overhead costs. These costs are essential to produce a product but cannot be specifically associated with a certain product. Overhead includes such items as taxes and insurance. All overhead costs must be recovered for the long term survival of the organization.

         Generally accepted accounting rules allow companies to use direct labor to allocate overhead expenses to the production process. [39]
      2. To simplify the evaluation of inventory [40].
3. What are the techniques available to compute direct labor standard hours?

a. Processes with a single operator
1. A formal time study

   A formal time study was the most common method based on the experience of the author to obtain a standard time and the most accurate for most assemblies, being very accurate for any cyclical assembly. A cyclical assembly is an assembly that repeats the same cycle each time the assembly is completed. See Glossary for a step-by-step method that will result in a successful formal time study.

   Prior to the initiation of the formal time study, each process must be evaluated for possible improvement.
2. For an assembly line with multiple operators

Methods improvement on an assembly line with multiple operators is similar to that of methods improvement with a simple operator but the first operation to be evaluated must begin with the bottleneck or the slowest operation on the line. Similar to the slowest car in a lane, no one can pass it until it is all clear. Improving the productivity of the operator before the bottleneck will only increase the work in process between that workstation and the bottleneck.

The first question that must be addressed is the identification of the bottleneck. If there are existing time studies for each operator on the assembly line, then the bottleneck operation can easily be selected since it is the slowest operation on the line. The simplest method of determining this is to list each operation in sequential order in one column and in the adjacent column list the normal elemental times for that element. Then several options exist, the simplest would be to sort the column containing the normal times in order, be sure that the file is saved before the sort order is executed. Then the slowest operation appears either at the top or the bottom depending on how the column was sorted. In the unlikely event there are two bottleneck operations on the same line, then methods improvement techniques must be applied to both. It does not matter which one you begin with since it must be used with both.

If there are no existing time studies available, then two possible methods exist. The first is to make several random observations of the line to determine if there is an accumulation of work in process at a given workstation. Several observations need to be made to confirm that the workstation is in fact the bottleneck.

After the bottleneck has been verified, then the method is the same as methods improvement for a single operator for which the author recommends a formal time study. See the Glossary for details.

2. Work sampling – frequently referred to as Activity Sampling or Ratio Delay Study is a statistical method first devised by L.H.S. Tippet in 1934. The objectives of Work sampling have two objectives:
   a. To measure activities and delays while a man is working and percentage not working.
   b. To measure manual tasks that can be used under certain circumstances to establish time standards for an operation.

The theory is based on statistics and states that the percentage of observations recorded on an operation in any state is a reliable estimate of the percentage time the operation is in that given that a sufficient number of random are taken at random [41].

To have an observation of working, the operator must be performing tasks that will add value to the product.

No conclusions can be made from the percentage of not working except that the operator is not performing tasks that will add value to the product.

c. Getting input from employees

One task the author had when installing a new computer system was to enter the routing times for items manufactured in a furniture plant. These items consisted of fully upholstered chairs and sofas with numerous options. The operation contained a machine room that began with raw lumber and included all needed operations to convert the lumber into a finished product. Different species of wood were used with consistent manufacturing times but wastes varied by species resulting in varying manufacturing direct labor and material costs.

Due to time constraints, the author asked several experienced employees in the upholster department the length of time required to upholster a certain chair. These times with allowances were inserted into the routing time for those chairs. The same process was used to obtain time for the spring-up department which inserts springs and the inner workings of the chair. These items are added immediately prior to the chair being upholstered. The same process for obtaining times was used for the hand sanding departments, and the cutting and sewing departments.

In the opinion of the author this is not the most reliable means of collecting information that will be used for the purpose of time of manufacturing and costing your product. The reasons are that although the author used three different estimates, they are still estimates, if you are to be judged or held accountable to a number, then you will give a relatively high number, one that you can easily achieve. The negative effects of this method are that the use of MRP will result in inventory being shipped too soon, and the direct cost of manufacturing can be grossly overstated.

So given all these negative reasons, why was it done? There were two engineers working full time on this project, there were at least 100K different routings and bills of materials and the company was under a time constraint – Y2K. Adding more engineers would have only compounded the problem since the products were unique. The process could not have begun earlier because the company had been recently purchased and the acquiring company utilized a different computer system.

The 2K in the Y2K problem refers to the year 2000. The reason that it was a problem was that at one time data storage was expensive so the amount of data stored was minimized. Only the last two digits of the year was stored. For example the year 1988 would be stored as 88. As the year 2000 approached, programmers realized that there would be a problem with various applications. For example, banks when paying or charging interest based on time, subtract the current time from the date of the loan or deposit to determine the amount. It was not known what would occur when the year was 00. This problem caused concern with the airline industry and the government as well. Estimates to address the problem ranged from 300 to 600 billion dollars. General Motors alone suggested an estimate of 585 billion USD whereas CitiCorp gave as its estimate 600 billion. The problem actually appeared, was addressed, and had little effect due to preparation and appropriate action [42].

   d. Group timing techniques – it is a work sampling technique that can be
      applied to numerous activities simultaneously so one observer can make an
      elemental time study that is very detailed for a short interval work sampling
      procedure for as many as 15 operators and as few as two. Operator rating
      is not considered due to the continuous observations being made at fixed
      predetermined intervals which tends to level the rating at 100% [43].

Each of these motions is given a predetermined standard time value or a previously fixed time value in such a way that the individual motions are used for total time. These times are summed to yield the normal time to perform a designated task.

The use of a default time system requires that the measurement is divided into basic movements/motions, as required by the particular system. All programs must be followed exactly by their own particular rules and procedures.

A predetermined time standard is one of many tools that can be used to establish standard time for the performance of a task. This tool uses times established for basic human motions depending on the motion itself and work conditions under which these motions occurred to develop the time for a task.

The major advantage of predetermined time standards (PTS) is that it allows the standard time for an activity/job to be predetermined if the all the motions are identified in detail. Through the workplace design and description of the method used, one can find out how long an operation would take beforehand.

The roots of PTS can be traced to the "therblig" as Gilbreth did in the chronocyclography, and the first attempt to establish time by motion length. Therbligs were used in all PTS s as fundamental motion units. Therbligs is simply Gilbreth spelled backwards.

A primary disadvantage of the use of a PTS system is that there is no decomposition of operations into elemental descriptions that allow the practitioner to determine distinct starting and stopping points and all tasks the operator performs between the two to seek improvement opportunities.

A formal time study (see Glossary for detailed instructions) is the method recommended by author based on his experience and the necessity of balancing assembly lines. The detail of each task that is delineated in the elemental descriptions by the time study practitioner while performing the studies provides the information needed to reduce the unforced idle time of operators and to an increase in output from the line with no increase in input. The provision of greater detail in the elemental descriptions enables the practitioner to achieve better results.

Based on the experience of the author, in any sequence of operations a constraint or bottleneck will exist which will limit the output that the sequence of operations can produce. The sequence of operations may occur on the manufacturing production line, in the system that converts a customer order into a manufactured and delivered product, and in various everyday life situations, as the checkout process at a grocery store.

    e. Does the operator have unforced idle time due to the quality of raw material that does not meet in coming specifications or a lack of work in process?

If the unforced idle time is the result of raw material that does not meet the minimal criteria, then the QA department must verify that the minimal criteria are critical to meeting the quality specifications of the final product, evaluate the current method used for inspection to determine its accuracy and ease of use and the training of the inspector. Any time lost due to a vendor providing materials that do not meet specifications should be recorded and the unmet specifications to enable the company to be compensated for the additional cost incurred. This information needs to be maintained for a vendor review for a possible vendor change.

If the unforced idle time results from an operator not having material on which to work, then each task in the operation must be studied to determine which tasks can be assigned to other operators.

Depending upon the frequency of lost production time due to quality deviations from specifications, the existing vendor may need to be evaluated for a potential change. All quality deviations should be observed during the initial inspections and the defects returned to the vendor. The frequency and severity of these occurrences should be documented and reflect in the history of the vendor

f. Does the operator have unforced idle time due to lack of or unclear communication?

Management must insure that all communication is clear and understood by all employees and employ whatever techniques needed.

g. Does the operator have less than desired performance due to the need for training?

Training should occur as needed to insure that the operator can operate at maximum performance with minimum waste and supervision. This is one of the many reasons that performances must be constantly monitored. The amount and type of training provided must be modified as the situation dictates.

h. Does the operator have less than desired performance due to other reasons?

There are numerous reasons that affect performance. Any that affect performance not related to work should be brought through appropriate channels and eventually ending in the department of Human Relations.

On a manufacturing production line consisting of sequential operator tasks, the constraint or bottleneck will result from the inequitable distribution of time required for each operator to perform his designated tasks. Possible solutions include a reallocation of tasks from the bottleneck operation to operators whose operations require less time than the bottleneck operation, the addition of an additional workstation which will necessitate a reassignment of all tasks, and cross-training of the operators who perform tasks before and after the bottleneck operation. Cross-training of each operator to perform several tasks would reduce the amount of time that operators are unproductive.

The reasons to balance an assembly line are to:

1. Increase productivity which will reduce direct labor costs and increased throughput at a minimum
2. Increase the capacity of the line which may result in potential increased sales

As an example to illustrate the methodology of reducing a bottleneck in an assembly line, Table 3.1 represents an assembly operation on which five operators work performing various tasks. Time studies were taken and shown in Table 3.1. Just as an automobile traveling slower than other vehicles is a bottleneck in the lane in which the slow automobile is traveling, the operator who requires the most time to perform his tasks is the slow driver who is restricting the flow of items being processed on the line. The mean times in hours in Table 3.1 are the results of formal time studies. The capacity of each operation represents the maximum number that the operator can produce per hour. The capacity is the reciprocal of the mean production time.

TABLE 3.1

Mean times per operation on an assembly line

Iteration 1	Opr 1	Opr 2	Opr 3	Opr 4	Opr 5
Mean average time (hrs)	0.06	0.04	0.05	0.07	0.03
Capacity per hour	16.67	25.00	20.00	14.29	33.33

TABLE 3.2

Computation of standard hours and capacity

Bottleneck operation		Opn 4
Mean of bottleneck operation		0.070
Allowances for PF&D (15%)	15%	0.105
Average rating or Operator 4	110	0.116
Standard hours per unit	5	0.578
Capacity of line per hour		8.66

As can be seen in Table 3.1, the bottleneck or limiting operation is operator 4 whose capacity is 14.29 units per hour. The average mean time must be converted to standard by the addition of allowances and the operator rating resulting in the normal output per hour. This calculation is shown in Table 3.2.

Operators who complete their assigned tasks in less time than the limiting or bottleneck operation will either experience idle time or produce items that will result in increasing work in inventory.

The standard direct labor hours for a unit produced on this line is a function of the mean time, the allowances for personal time, unavoidable delay and fatigue, and the rating of the operator.

Calculation of productivity of the line must be based on the constraint, operation four. This computation is seen in Table 3.2. The practitioner rated operator at a performance level of 110%, the company for this line has determined that the combined allowances for PF&D is 15%.

To calculate productivity, assume that the line produces 60 units in an eight-hour period. Productivity is the ratio of hours earned and actual hours worked to earn those hours. These computations are shown in Table 3.3.

The standard hours earned are the product of units produced (60), the specified time period (8 hours), and the standard hours per unit (0.578). The number of actual hours worked to earn 34.65 hours is determined by multiplying the actual crew size for that period (5) by the number of hours worked by each crew member.

To insure the reasonableness of the calculated productivity, 87%, compare the mean units produced per hour and with the maximum number per hour that could

TABLE 3.3
Calculation of productivity on an assembly line with bottleneck

Example		Std hrs earned
Line produces 60 units in 8 hrs	60	
Standard hrs earned		34.65
Actual hrs worked	8	40
Productivity		87%

TABLE 3.4
Effects of shifting tasks on an assembly line to improve productivity

	Opr 1	Opr 2	Opr 3	Opr 4	Opr 5
Mean average time (hrs)	0.05	0.04	0.05	0.06	0.04
Capacity per hour	20.00	25.00	20.00	16.67	25.00

be produced by the constraint. As can be seen in Table 3.2, the maximum is 8.66 units per hour, the average produced by the line during that period is the number produced (60) divided by the hours consumed (8) to yield an average of 7.50. Thus since the actual output per hour is less than the maximum that could be produced, the computed productivity should be less than 100%, which is (87%). An average productivity greater than the maximum (8.66) would result in a productivity greater than 100% and conversely.

To achieve improvement, the average number of units produced per hour must be increased. In order to accomplish this, the elemental descriptions documenting the tasks of each operator must be evaluated to determine which tasks can be shifted from the bottleneck operation to another operator. The purpose in shifting one or more tasks is to decrease the time needed in the bottleneck operation which causes another operation to become the bottleneck depending on the amount of time that was involved.

An example to demonstrate the effects of a shift in tasks from one operator to another. The practitioner after an analysis of the elemental descriptions realizes that one task currently performed by operator four could easily be performed by operator five. The mean time for that task is 0.01 hours. Results after moving the task can be seen in Table 3.4.

As can be seen in the above table, shifting the task from operator four to operator five did not result in a change in the constraint which remains operator number four. It did reduce the time that operator number four needed to complete his reduced workload 14.3%. This percent reduction in standard hours is the same due to the allowances for PF&D and the rating factor not changing. The results of this

TABLE 3.5

Effects of shifting tasks to improve line productivity

		Iteration 1	Iteration 2	Percent change
Bottleneck operation		Opn 4	Opn 4	
Average of bottleneck operation		7.0%	6.0%	−14.3%
Allowances for PF&D (15%)	15.0%	10.5%	9.0%	
Average rating or operator 1	110 %	11.6%	9.9%	
Standard hrs for process		57.8%	49.5%	−14.3%
Capacity of line per hour		865.8%	1010.1%	16.7%

TABLE 3.6

Effects of shifting tasks on the bottleneck operation

	Opr 1	Opr 2	Opr 3	Opr 4	Opr 5
Mean average time (hrs)	0.046	0.04	0.044	0.05	0.04
Capacity per hour	21.74	25.00	22.73	20.00	25.00

calculation are shown in Table 3.5. The standard labor costs for this process will experience the same percentage reduction.

A small change in the process time for operator four from 0.07 hours to 0.06 hours or a reduction of 14.3% results in a larger increase in capacity, 16.7% due to the number of operators (five).

An examination of Table 3.4 reveals another possible opportunity for improvement. Operation four remains the bottleneck. The practitioner examines the elemental descriptions for potential tasks in operation four that can be transferred to other operations. Since operations one and three have the same mean time (0.05 hours), tasks from operation four, the bottleneck, must be transferred to both operations to have the maximum effect. The transfer of a task to another operator requires two processes. In this situation one to be transferred to operation one and the other to operation three. The maximum amount of time that can be transferred is the difference between the mean time of the operation from which tasks are to be transferred to the operation or operations that will receive the new tasks. As seen in Table 3.4 this difference is 0.01 hours. Since the transfer of tasks will be distributed to two operations, the maximum that can be transferred is 0.005 hours. An examination of the elemental descriptions reveals two tasks that can be transferred, one with a mean time of 0.004 hours and the other a mean time of 0.006. These changes can be seen in Table 3.6.

The constraining operation remains number four but the transfer of tasks from operation has resulted in an increase in the amount that the line can produce as

TABLE 3.7
Computation of standard hours
based in bottleneck operation

Bottleneck operation		Opn 4
Mean of bottleneck operation		0.050
Allowances for PF&D (15%)	15%	0.058
Average rating or operator 4	110	0.063
Standard hrs for process	5	0.316
Capacity of line per hour		15.8

TABLE 3.8
Revised direct labor standard hours

		Iteration 1	Iteration 2	Iteration 3	Percent change from iteration 1 to 2	Percent change from iteration 1 to 3
Bottleneck operation		Opn 4	Opn 4	Opn		
Average of bottleneck operation		0.070	0.060	0.050	−14.29%	−28.57%
Allowances for PF&D (15%)	115.0%	0.081	0.069	0.058		
Average rating or operator 1	110.0%	0.089	0.076	0.063		
Standard hrs for process	5	0.443	0.380	0.316	−14.29%	−28.57%
Capacity of line per hour		11.3	13.2	15.8	16.7%	40.00%

defined by the constraint from a maximum of 16.67 to 20.00 units per hour. The revised direct labor standard hours are shown in Table 3.7.

The constraining operation remains number four but the transfer of tasks from operation four to operations one and three has resulted in an increase in the amount that the line can produce as defined by the constraint from a maximum of 13.2 to 15.8 per hour. The revised direct labor standard hours are shown in Table 3.8.

Continued efforts to balance the assembly line have yielded an increased capacity of 40% from the initial results of 16.7%. These continued improvements led to a further reduction in the standard hours to complete the process from 14.29% to 28.57%. The reduction in the standard direct costs will also realize the same percentage decrease.

i. Is there a system in place to record production?

This is the first action item that must be addressed. Production must be recorded as in the routing file as units per hour. Based on the author's experience, the most accurate is electronic using bar codes. Depending on the use, scanning equipment on the line will provide the most accurate and recent information.

The use of bar codes will enable a change to be detected immediately whether positive or negative and provide feedback to management and to the individual operator as to progress to goals.

j. Who takes the measurements and what minimum qualifications are needed?

The actual measurements should be taken by a trained observer to insure the consistency of the information. The training should include a complete comprehension of the methodology to be used to collect the data, a basic knowledge of human factors specifically anthropometry and how its application can result in improved productivity and employee satisfaction, interpersonal skills, and emphasis on the ability to communicate verbally and orally.

k. How will waste or products that cannot be used for production be recorded?

Only usable products must be counted and the count has to be verified for accuracy to insure that shipment shortages do not occur and that payrolls are accurate.

l. How supportive should management be of the program?

This program as well as all programs must have the full and undivided support of all levels of management. This just does not mean posters and banners but real employee engagement.

m. How involved should the employees be in the establishment of the program?

Each employee should understand the computation of a rate, its derivation, and meaning. Each employee has the right to ask any questions since this affects their earnings and employment status.

The percentage of downtime that an operator incurs may indicate that a problem exists somewhere in the system. The percentage should be close to zero, anything greater than zero could indicate an imbalance which warrants further study, the greater the deviation from zero, the greater the problem. The downtime reports create another situation for the supervisor; he has to insure their accuracy and total the hours and report them to payroll since the operator earns incentive pay on the hours worked less the downtime hours. Regular pay is earned on downtime hours.

n. Will the skill level affect the study?

The skill level would be compensated for in the rating factor given to the operators by the practitioner. This is another reason to insure that studies are conducted by different practitioners on the same operators on different days and to validate results.

o. Does the operator have to wait on parts before he can start working?

If there is a delay that prevents the operator from immediately beginning to work on the assigned task, then the preceding operation must be studied to determine the case of the imbalance. The time needed to complete the preceding process should be somewhat greater than the following one to prevent the operator from having to wait on a component to begin work. The operator should not complete his assigned tasks too quickly since that will result in the succeeding operator being forced to wait until he completes the unit on which he is currently working. The current method used in installing labor standards?

p. Is the group to be studied represented by a union?

The only difference this will make is that after notifying the supervisor of the intent and the reason to take work measurement studies, the next person to notify is the person representing the union. Together if the representative desires, then the person selected for the study should be approached. This applies to each time a practitioner desires to take a time and motion study. The first person is the supervisor then the union representative. Should the person for some reason that you ask choose not to participate in the study, then thank him and ask another person. His reason for non-participation is confidential and personal [44].

C. CRITICAL QUESTIONS TO ASK FOR REVISION OF EXISTING DIRECT LABOR STANDARDS WITH SUGGESTED SOLUTIONS

1. What necessitated the need for change?

The requirement to adjust direct labor standard hours can result for numerous reasons

a. Was there a change in a raw material purchased?

This could result in the need to change a manufacturing process, a quality inspection procedure performed during manufacturing, receiving, or delivery of the final product. The change may affect packaging specifications including the time allocated for packaging. Service and repair parts can also be evaluated to determine all the effects that this different material will have on them.

b. Has there been a change in a manufacturing process?

A comparison of the current manufacturing process with the previous method using the detailed elemental descriptions will reveal any deviations between the two. Only those elements that have changed should be studied for possible revisions.

c. Has there been a change in the working environment that would necessitate a revision in the allowances for delay, personal time, and fatigue?

This revision could result from the location of machines changed, a change in a fixture, or a modification in the environment in which the operators perform their tasks. Any change would be detected during an evaluation of the differences in the elemental descriptions for the task from before the change and after the change became effective. Appropriate studies would then be made via an ECO.

d. Have all the changes that this change in direct labor standards been detailed in the ECO?

These changes will include the shippable product, service, and repair parts. The originator of the ECO is responsible for its accuracy.

e. What is the effectivity date of this change?

The material planning function will determine this date based on the severity of the need for the change. Changes affecting the safety of employees, the need to correct OSHA, or other violations including environmental must be immediate to prevent possible injuries to employees, manufacturing processes, equipment, or the environment. For other changes, the material planner will evaluate existing inventory levels, the length of time needed to obtain the new item, the usage of the requested change, e.g. a service or repair part. This date has several ramifications. First, it will effect direct labor costs of all the items contained in this operation and the existing schedule of these items. This date must be monitored closely.

f. Who will make the necessary changes and insure the accuracy of those changes?

The originator of the ECO is responsible for verifying that all areas that need to be changed are included in the ECO, effect of the change, and accuracy of the data. For example, should the newly purchased material be a different grade of lumber for a furniture company, how will this affect waste generated during each process, process time, incoming inspection procedures, and of those conducted during manufacturing, and the tare weight of all shippable products including service and repair parts. His responsibility also includes insuring the accuracy of the information.

2. Which of the available techniques for determining direct standard labor hours will be used to make the changes?

The same techniques applicable to determine initial direct labor standard hours can also be used to reevaluate existing rates. These are:

1. A formal time study
2. Work sampling
3. Getting input from employees
4. Group timing techniques
5. Predetermined motioned time systems

Since the goal of reevaluation is improvement, the technique recommended is a formal study. The elemental descriptions provide opportunities for improvement as well as tools to evaluate proposed changes to insure that the proposed changes are cost effective. This recommendation is applicable to processes involving a single or multiple operators.

Examples of successes and failures are discussed in detail. The critical questions that were asked prior to the beginning of studies are presented to demonstrate that the use of critical thinking will increase the probability of a successful implementation. A list of questions that should have been asked and answered in the application that failed will be listed to strengthen the need for critical thinking.

D. EXAMPLES OF SUCCESSFUL AND UNSUCCESSFUL REVISIONS

  1. Examples of a successful revision
     a. An example of a successful application of a revision of existing direct labor standards occurred in a furniture company for which the author worked which produced dinette furniture that could be purchased upholstered in either fabric or vinyl. The furniture was metal framed with some models having arms and other components made of wood that was stained. The direct labor standard hours for the sewing operation of seats and backs were based on three sizes, small, medium, and large initially. There were six standards for all the different items sewn regardless of the type of material sewn, the shape of the item sewn, or if the cover contained a welt cord. Back covers were sewn on three sides, the bottom was open to facilitate the upholstery operation. Some back covers were sewn with welt as were some seats. The seats were sewn around the circumference, some of which contained a welt cord that was sewn into the seat. Installation of the welt cord was accomplished by the feeding of the welt cord through a fixture that wrapped the welt cord or piping into a one-inch strip of fabric identical to the one being sewn.

Observations made as sewers worked on different shaped patterns that the sewing time per inch of covers with sharp curves would be greater than those with a less curvature.

The author could not find any information that documented the derivation of the existing standard direct labors. The standard for sewing a small back or seat was identical for fabric or vinyl with or without welt cords and regardless of shape.

This company, as do most, experienced changes in management and engineering. Due to the lack of documentation as to the differences between, small, medium, and large, there was no tool that could be used to determine standard times for new product introductions.

Management decided to devote the necessary resources to accomplish this task.

The critical questions that were asked prior to the initiation of studies:

1. What level of accuracy is desired and how is that achieved?
2. What effect on the standard time would the addition of a welt cord have?
3. Will the rate for sewing fabric and vinyl be different?
4. How will sewing times for curved shapes differ from those sewing straight ones?
5. How can a tool be established to determine sewing standards for future products?
6. How can these standards be maintained to prevent future obsolescence?

Management decided that due to the confusion concerning the existing rate, the level of accuracy would be 95%. The plant and industrial engineering management was in full agreement as to this level of accuracy. The last factor in determining the normal time for an operation is the rating of the operator assigned by the practitioner who is performing the actual time study.

To insure consistency among the practitioners, examinations of the same tasks were performed and observed by the two different practitioners and studies did not begin until there was at least a 95% agreement among the operators. The third practitioner was used to randomly verify the accuracy level of the studies.

The supervisor provided each sewer a bundle of covers to be sewn which was tied together with identifying information based on the availability of the operator and the need of the schedule. The sewer would scan the covers into the computer when completed and put them on the conveyor belt to enable the counter at the end of the line to verify the quantity and turn the covers as needed. Some covers, as backs, were sewn inside out and needed to be reversed before the upholstery operation could begin. The turning operation was a step in the inspection process.

The new standards were installed. Prior to installation management decided to provide additional training to any employee who did not attain a 90% level of productivity.

Results of the studies revealed that the sewing time per minute did not vary statistically between fabric and vinyl, curved or straight sections, and with or without welt cord.

A standard data sheet was developed that contained the date the sheet was completed, the person who completed the sheet, the style number, pattern number, and a detailed illustration of the finished product.

b. Another successful example

While employed with a paper converting company, the company acquired a smaller one that specialized in producing paper tubes for the paper industry. This acquired company had five smaller plants that supplied a major producer of paperboard used for the manufacture of corrugated containers and other packaging material comprised of paperboard. The tubes were larger than the range produced by the purchasing company; the typical inside diameter was six inches, the thickness was 0.500 inches with a length of 144 inches or 12 feet. These tubes were used to package the large rolls of paperboard which was 12 feet wide.

During visits to each of these newly acquired plants, the author performed formal studies and then compared the results with the production standards that he had previously established with larger companies. The labor-derived standards for each item produced in the smaller plants were less than the standard for the identical item produced in the larger plant. The differences varied as much as 20%.

The author with the assistance of all levels of management began evaluating the methods used in the smaller plants and compared them with the methods used in the larger plants to the reasons for the difference. Due to the detail provided in the elemental descriptions of the formal studies taken by the author in the larger plants' management, he was able to determine the changes that needed to be made in the methods used in the smaller plants and as a result, productivity in the smaller plants increased by as much as 20%.

2. An example of an unsuccessful installation

An example of an installation that did install direct labor standards but due to the limited amount of time provided the outside consulting company to establish the standards and a monitoring system, the outside consulting company was not able to make necessary improvements that would have resulted in lower direct labor costs but would have simplified the task of establishing and monitoring the rates considerably.

To illustrate the missed opportunities in this plant, several examples will be discussed in detail. The existing process on which direct labor standards were established will be presented. Then critical questions that were not asked will be discussed in detail to indicate the possibilities for improvement and the effects that implementation of those improvements would have had on the discussed processes.

One of the products produced at the plant was aluminum mini-blinds. The blinds were painted with liquid paint and were available in over 200 different colors. The minimum width available was 9 inches and the maximum was 96 inches; the minimum length was 24 inches and the maximum was 96 inches.

The manufacturing process consisted of numerous distinct steps:

a. Cutting the top and bottom rails, punching the top rails to accept the mechanisms that controlled the positioning of the blinds. An inventory of top and bottom rails in all colors, over 200 were available during this period, was maintained at each of the five workstations that produced the top and bottom rails.

b. Installing the controlling mechanisms into the top rail, cutting and installing the cords and ladders, the ropes enabled the raising and lowering of the blinds, the ladders which resembled a ladder, was the length of the blind and held an individual blind, both were the same color as the blinds. There were five workstations that performed this task. An inventory of cords and ladders for each of all the colors was maintained at each cutting which also installed and a required number of individual blinds into the ladders of each workstation.

c. Inspection and installation of the ropes through each individual blind.

d. Inspection, positioning, and installing the bottom rail to meet the length specifications.

e. Wrapping the completed blind with protective bubble wrap for packaging.

f. Placing the blind into a corrugated container, add protective packing material as needed, cut the container to the correct length and insure closure and add the shipping label.

The implementation was not as successful as it could have been due to the fact that the studies were taken on the existing operations and there was no effort made to improve the existing operations.

The following critical questions were not asked but would have resulted in an improved outcome.

1. Were existing methods evaluated to determine to what extent methods improvements could have been implemented prior to the initiation of the studies?

    This is the most critical aspect of any project. Tasks that are not needed must be eliminated. To quote Dr. Peter Drucker, "There is nothing so useless as doing efficiently that which should not be done at all" [45].

2. How much involvement in the studies should the employees have had?

    The employees are the key to the success of the company and any program. The employees should have been fully engaged from the beginning of the project to insure the accuracy of the studies.

    Employees need to explain in detail the methodology of determining the rates, interpretation of the meaning of the rates, the reasons for establishing rates, the goal the company hopes to achieve as a result, and the efforts the company will make to those who need training to meet the minimum levels of output needed to remain employed.

3. How much involvement in the studies should the plant engineer have been to enable him to answer questions after the consulting company leaves.

    The plant engineer should have been intricately involved in the establishment of all rates. After the consultants leave it will be the responsibility of the plant engineer to maintain these systems which cannot be done without a clear understanding of how they were established and to answer any question the operator may have.

4. How were costs for the blinds established?

    The author was unable to determine a methodology that was used to calculate the costs. Prices were based on the square feet of the material used which would result in some product being less profitable than others, for example, blinds that were at the extreme range of products produced, as blinds that exceeded 10 feet in length and those that were less than 18 inches wide. During the production of blinds that met that criteria, either too long or too narrow, the production speeds had to be reduced to meet the specifications of additional material handling that was required due to the cumbersomeness of the blinds.

The author was not able to determine strategies to increase profits. Scheduling of particular orders through the various manufacturing processes was not possible due to the lack of standard labor hours.

5. What was the distribution of orders by color?

Forty percent of all orders consisted of two colors equally distributed between white and alabaster.

The author suggested that 40% of the five work centers be allocated to those colors, one for white and one for alabaster. The remaining three would be used to manufacture all the other colors that were available. This would reduce inventory at two workstations and the amount of touch up that resulted from the interaction of the liquid painted rails.

6. What was the percentage of all the available sizes that the blinds studied to determine the direct labor standards?

This question is important due to extreme variability of the sizes that are manufactured. Studies must include at least 95%, which is two standard deviations away from the mean for the data to be meaningful. Each machine has a limitation that results in a production rate or percentage of defects that deviates from the mean. Data must be acquired for the extreme points in the sizes available since it is at these extreme points that the greatest deviation from the mean will occur. Costs must reflect these endpoints to prevent the sales of items below cost.

7. A standard cost system using standard direct labor costs, the material costs which would vary with the intended purpose and location, and the overhead allocation. The material costs would include waste generated during the various manufacturing processes.

4 Integrating Measurement into the Establishment of New Direct Labor Standards or Revision of Existing Direct Labor Standards

A. BRIEF EXPLANATION FOR INTEGRATING MEASUREMENT INTO THE ESTABLISHMENT OF NEW DIRECT LABOR STANDARDS

Measurement is essential to insure that the initial goals established are being achieved, that the initial goals are obtaining the desired results or the need to modify those initial goals and that continuous improvement in those goals are being realized. The units of measurement employed to obtain the initial status of the process being evaluated must be the same as those used to evaluate the process after implementation of the proposed modifications.

B. CRITICAL QUESTIONS TO ASK FOR THE INTEGRATION OF MEASUREMENT INTO RECENTLY INSTALLED DIRECT LABOR STANDARDS WITH SUGGESTED SOLUTIONS

1. WHAT TECHNIQUES WERE USED TO COLLECT THE INFORMATION?

The technique employed and the level of accuracy obtained will determine the most appropriate application as discussed in detail below
The available techniques as discussed in Chapter 3 are:

- a. A formal time study
- b. Work sampling
- c. Getting input from employees
- d. Group timing techniques
- e. Predetermined motioned time systems

DOI: 10.1201/9781003153412-4

The formal time study produces more accurate results and due to the level of documentation, greater opportunities for improvement opportunities. See Glossary for detailed instructions. The results of a formal time study can be used in any application but particularly those that demand the greatest accuracy and improvement opportunities which include the computation of direct standard labor costs, incentive systems, and process routing information because of the serious effects that inaccuracies can cause. Accurate direct standard labor costs effect profitability, incentive systems effect employee income, and routing information affect scheduling.

The use of work sampling techniques should be limited to those situations that require an approximate value as the number of indirect employees required in the receiving department needed to achieve certain goals. These studies should always be duplicated periodically to validate the result.

The technique of obtaining information from employees should only be used when the information is needed sooner than using a more accurate method would provide. In the opinion of the author getting information in this manner should only be employed as a last resort.

Group timing techniques can provide accurate information but the degree of accuracy is directly related to the design of the study and the number of repetitions.

The accuracy of predetermined motioned time systems depends on the ability of the observer to decompose each task into individual motions and how well an existing operator's particular motion can be aligned with one from the table from which predetermined values are retrieved.

The method used by the author is formal time studies because it enables the practitioner to:

a. interact with the various operator which provides the practitioner to discuss improvement opportunities.
b. observe for bottlenecks as they occur for their elimination or reduction.
c. verify that the operation is critical and cannot be eliminated.

2.   WHAT ARE SOME APPLICATIONS FOR DIRECT LABOR STANDARDS?

a. The determination of standard direct labor costs.
b. A basis for the allocation of indirect labor costs.
c. The computation of direct standard labor hours includes an allowance for personal time, fatigue, and any additional allowances needed for hazardous conditions.

3.   OF WHAT SIGNIFICANCE IS THE LEVEL OF ACCURACY?

The level of accuracy of the results of the studies depends on the intended application. Applications involving direct labor as determining the manufactured cost of a product, the value of inventory, and the earnings of employees must have the greatest accuracy, the author has sought a 95% level or higher. Other applications as determining the workload of indirect labor using work sampling techniques should have an accuracy level of at least 70%. Recommended solutions from these applications should be verified by additional studies before the solutions are implemented.

4. HOW DOES THE MEASUREMENT RELATE TO DIFFERENT CREW SIZES?

The labor standard will be based on the crew size that yields the maximum production with the minimum waste while enabling the crew to maintain a comfortable work pace. Deviations from the standard crew size will effect productivity. For example, if a crew of five earns six standard hours that normally should be earned by a crew of four, the productivity would be a maximum of 80%. If the machine had a crew of three and earned six standard hours, it could earn a maximum of 75%, but theoretically in the long run it would be difficult for three people to do the work of four if the rates are set correctly.

In the example discussed above concerning the company manufacturing window coverings (page 72), the failure to ask and answer the critical questions resulted in the worst possible outcome – the closure of the business and the loss of over 400 jobs.

5. WHAT IS THE DEFINITION OF A UNIT?

A unit is defined as whatever the production crew is working on at the time the studies are being conducted to establish the labor standard. The unit could be a component, a sub-assembly or the entire assembly.

6. WHAT WAS THE GOAL OF THESE APPLICATIONS?

The goal of every application must be improvement. The percentage of improvement will vary but it is critical to the long-term existence of the organization that this percentage is monitored with the distinct purpose of continued improvement.

7. WHAT INFORMATION WILL BE MONITORED LONGITUDINALLY
REASON TO TRACK AND EVALUATE PROGRESS?

The decision of what information to track must depend on the criticality and the priority of the requirement when the installation occurs, the situation of the organization at the time of installation, and the resources available. A company that is struggling will elect that information essential to its immediate and long-term survival.

The information that is initially selected must be continually monitored for change to provide current needs for the needs of the organization.

8. WHAT BENEFITS WILL RESULT AS A RESULT?

The goal of installing a measurement system is improvement of the particular matric. The percentage of improvement can then be determined and used to prioritize efforts to attain the goals of the

9. WHAT IS THE MOST BENEFICIAL PLACE TO SEEK IMPROVEMENT?

Based on the many years of experience of the author, the goal of the industrial engineer is to seek solutions that provide the most benefit with the least costs. Elimination of a bottleneck will satisfy that goal. Every operation has a constraint or a bottleneck and production is limited by that constraint. Continue to eliminate bottlenecks until the process is no longer cost effective. The bottleneck can be a function other than production.

C. CRITICAL QUESTIONS TO ASK FOR THE INTEGRATION OF MEASUREMENT INTO EXISTING DIRECT LABOR STANDARDS WITH SUGGESTED SOLUTIONS

1. WHAT WAS THE REASON FOR THE INITIATION OF THE NEW STUDIES?

Changes must have occurred to initiate studies to evaluate existing standards. Possible changes include:

a. A different vendor of an existing purchased item.

All operations that involve this item need to be investigated to determine if this item from the new vendor requires a deviation from that currently employed. These deviations, the degree of deviations, the reasons for them as well as all the operations affected must be explicitly explained in the approved engineering change order that requested the new studies.

b. A change in the specifications for the product or assembly.

As with a vendor change all operations that involve this item need to be investigated to determine what changes resulted from this specification change and their extent. These changes could occur in the receiving, manufacturing process or packaging processes, or all of them. A comparison of existing documentation reveals differences. It is for this reason that detailed documentation for each process exists. Studies can then be focused on the differences. As in a vendor change, the deviations, the degree of deviations, the reasons for them as well as all the operations affected must be explicitly explained in the engineering change order that requested the new studies prior to its approval.

c. A change in a component used in the manufacturing process.

Each usage of this new or reconfigured component must be studied to determine the operations effected by this new component and its effects measured, justified, and documented.

d. A change in the required skills of an operator.

The change in skills needed to perform a task may affect the quality or the quantity produced by that operator.

e. A change in the work environment.

The work environment must be documented in the elemental descriptions and elsewhere as the studies to establish the initial direct labor standards were taken. This original documentation is the basis to recognize that a change has occurred. The new ECO written to request the new studies must include the difference between the two environments, the reason for the change, and its effects on operators affected by the change. The change could include the requirement that two operators are now needed to transport an item due to a weight increase in the item, a different manufacturing process, as the replacement of a liquid paint line to an electrostatic (a powder cost) system, or the replacement of equipment. These changes may affect the time for a process, the required PPE, or any of the allowances.

2. WHAT IS THE EFFECTIVITY DATE OF THE CHANGE FOR EITHER AN ITEM WITH A CHANGED SPECIFICATION OR A COMPONENT USED IN THE MANUFACTURING PROCESS?

The material planning function will determine this date based on the severity of the need for the change. Changes affecting the safety of employees, the need to correct OSHA or other violations including environmental must be immediate to prevent possible injuries to employees, manufacturing processes, equipment, or the environment. For other changes, the material planner will evaluate existing inventory levels, the length of time needed to obtain the new item, the usage of the requested change, e.g. a service or repair part. This date has several ramifications. First, it will effect direct labor costs of all the items contained in this operation and the existing schedule of these items. This date must be monitored closely due to unexpected changes in orders, delivery time, or other potential interruptions.

3. A CHANGE IN A COMPONENT USED IN THE MANUFACTURING PROCESS

a. For a vendor change, the planner must consider the existing inventory quantity and the delivery time from the new vendor at a minimum.

b. For a change in the specifications for the product or assembly or a component used in the manufacturing process the needed changes must be stated in the ECO. The planner will incorporate the existing inventory quantity and the availability date of the item with the revised specifications into the decision. The planner must include the available repair and service parts and their order frequency.

D. EXAMPLES OF SUCCESSFUL AND UNSUCCESSFUL INTEGRATION OF MEASUREMENT INTO DIRECT LABOR STANDARDS

1. AN EXAMPLE OF A SUCCESSFUL IMPLEMENTATION

As discussed in the example in Chapter 3, the application of the direct labor standards was successful in the smaller plants due to several factors, the degree of detail in the elemental descriptions that enable management to discern any differences between processes, the commitment of management to minimize these differences by allocations of resources for maintenance and training.

The sales prices of these spirally wound paper tubes are per M-tubes based on four variables, the grade of paper, the ID of the tube, the wall thickness of the tube, and the length of the tube. Formal time studies revealed that the limitation of the speed of winding the tube was the time that the saw that cut the tubes had to return to its original position to reengage for another cut. Thus this was a bottleneck for the speed with which spirally wound tubes could be produced. Studies were taken focusing on length and the result was an upcharge for the shorter lengths.

To determine the sales price per M-tubes, the salesman would extract from a specific matrix the sales price per M-linear inches and multiply by the length of the proposed tube to arrive at a sales price per M-tube and add upcharge per M-tube as needed. Latex, a sticky material used to increase adherence to the tube as material was wound onto the tube, was an example. Another example was the addition of a flap which was used to tuck the material to be wound. Each tube diameter in increment of ⅛ of an inch was represented by a different matrix. The columns listed the various wall thickness available and prices per thousand linear inches occupied the values in each cell.

Direct labor standards were established and presented in the same format to enable the establishment of a standard cost system. The system was based on standard direct labor costs, the material costs which would vary with the intended purpose and location, and the overhead allocation. A productivity that deviated from 100% would result in a direct labor variance. If productivity exceeds 100%, the variance was positive, conversely if less than 100%. The material costs would include waste generated during the various manufacturing processes calculated using the ID of the tube, the wall thickness, and the length. Numerous previous studies that were conducted to determine if the aper had any effect on productivity concluded that there was no effect on the speed. This was due to the process of manufacturing spirally wound paper tubes.

However, this is not always the situation. Each manufacturing process must be evaluated separately. For example, the manufacture of paperboards necessitates a slower production speed for the higher grades of paper resulting in greater direct labor costs per ton than the lower grades. This information was used to save the paperboard manufacturer money in utility costs by shifting the production of the higher grades to different shifts to take advantage of off-peak costs of utilities.

The direct labor standards that the author had established began to be used to determine productivity for all the plants. This enabled comparisons of identical items among all the plants and was an incentive for plants with a lower productivity percentage to improve.

Productivity percentages were calculated at the end of the shift which is not beneficial to improve production as it occurs. The production standards were in thousand linear inches and the production units were individual tubes. A measurement must be meaningful and provides assistance to the operator to assess the current status of production. In order for the operator to determine his current level of productivity, the standard must be stated in units that can be easily calculated. The operator could not count thousands of inches of tube per hour but the number of tubes per minute could be easily determined. Thus each order would have the standard tubes per minute on the work order to enable a real-time calculation of productivity so that adjustments could be made as needed.

The adaption of identical standards and the inclusion of the standard tubes per minute resulted in increased productivity for all plants, some as much as 20%.

2.   AN EXAMPLE OF AN UNSUCCESSFUL IMPLEMENTATION
OF PRODUCTION STANDARDS

The author developed direct labor standards in a plant that manufactured polyester fabric that was slit into narrower widths that would then be wound into various lengths for processing into casts to assist in the mending of broken arms and legs, etc. The plant that converted the polyester fabric into casts was located in a distant location.

Processing the polyester fabric consisted of weaving the yard into rolls, then slitting the rolls into three different widths, and then rolling each width into designated lengths that were shipped to the facility that processed these lengths into casts.

Prior to the initiation of the studies the author was assured that the existing computer system contained the necessary software that could employ the direct labor standards to determine productivity and for scheduling. After the direct labor standards for all operations had been determined and validated manually, it was determined that the existing software could not perform the necessary calculations.

This improvement effort could not be fully implemented due to the failure of the author to confirm that the existing computer system could use the standards to calculate productivity and to schedule orders through the plant. Since these proposed functions had never been done on the existing computer system, the system should have had a test run with some estimated data to verify that it would perform as required. The author was advised that the needed software could not be purchased at that time due to the costs. As a result, the resources consumed to develop the standard data were essentially a waste. The data could be used at a later date after the installation of the software but studies would have to be made to verify the accuracy and make necessary adjustments for changes.

The reason that this attempt was not successful was that the author did not confirm the information provided. The author assumed which should never be done. As President Reagan stated on numerous occasions, "trust but verify".

5 Integrating Methods Improvement for a Possible Revision of an Existing or the Installation of a New Incentive System

A. BRIEF EXPLANATION FOR NEED OF METHODS IMPROVEMENT PRIOR TO THE INSTALLATION OF A NEW INCENTIVE SYSTEM

Minor improvements in procedures, the working environment, and the application of the learning curve due to the ever-increasing repetitions of the task may result in improvements that need to be documented, which will require a detailed ECO to insure that modifications in affected documentation will occur. Unless modifications are made, these minor improvements will cause an increase in direct labor costs due to an increase in productivity. The ECO will examine these minor changes for their effect on direct labor standards.

B. CRITICAL QUESTIONS TO ASK BEFORE INSTALLING A NEW INCENTIVE SYSTEM

1. Have any changes in manufacturing processes occurred since the time studies are taken to establish the standard direct labor hours?

   Current methods established must be examined to insure that no changes in methods used when time studies to establish direct labor standards were taken have changed. A detailed study of written elemental descriptions that accompany as the time studies being taken will reveal any changes in processes, procedures, and materials. If any changes have occurred, then new studies of the changes must be taken and the results made in direct labor standards before individual incentive systems can be implemented.

2. How supportive are all levels of management?

   Management must fully support the system and demonstrate that support by providing the necessary resources to insure that the system is accurate, provides everyone equal opportunities to earn additional monies, and enables the employees in the system to easily calculate the extra monies earned from the incentive system.

DOI: 10.1201/9781003153412-5

3. Did any employee experience an incentive system with previous employers?

These employees will have an understanding of the reasons for the system and the mechanics of the system. They will be a benefit as the studies are initiated since they will be able to explain the system and answer questions from other employees.

4. How will temporary periods that operators cannot work due to lack of materials be treated?

This time, often referred to as downtime, is frequently maintained by each operator separately and verified by the supervisor for accuracy. Operators are normally paid at their normal base rate during downtime.

5. How will an operator be paid if he is transferred to another department due to lack of work in his normal department?

Based on the experience of the author, he is paid the higher of the two base rates, the one at the job to which he was transferred or his old one. He is being transferred to this position at the request of and for the benefit of the company and should not suffer financial losses.

6. What level of involvement should the employees have who will be in the system?

It is crucial that these employees are completely involved to insure that they are knowledgeable concerning the methodology of establishing the rates, calculating productivity and earnings. This participation begins when the management makes the decision to implement the system. An explanation of the studies to be taken, the methods that will be used, the people who will take the studies, the calculations that will be used to obtain the results, the meaning of the results, and the validation methods that will be used to validate the accuracy of the results.

7. What will occur for employees who do not meet incentive goals?

These employees will receive the necessary training needed to improve their skills to enable them to earn incentive pay within the time period established by management. Should those employees not be successful, then other positions within the organization should be investigated for possible transfer.

8. Is the group to be studied represented by a union?

See question 8 below

C. CRITICAL QUESTIONS TO ASK BEFORE INITIATING STUDIES TO MODIFY AN EXISTING INDIVIDUAL INCENTIVE SYSTEM WITH SUGGESTED SOLUTIONS

BRIEF EXPLANATION FOR THE NEED FOR METHODS IMPROVEMENT FOR AN EXISTING SYSTEM

1. What changes in manufacturing processes, methods of reporting, or data collection have occurred since the time studies that were used to establish the standard direct labor hours?

Current methods must be examined in detail to insure that no changes in methods used when these time studies were taken have changed. A detailed study of elemental descriptions written as the time studies were being taken will reveal any changes in processes, procedures, and materials. A change is the only justification for initiating new studies that could result in revision of existing incentive rates. As a result, the criticality of detailed accurate documentation cannot be overemphasized. The ECO process must be stringently followed.

2. Has there been a change in incentive earnings over the last few years?

Due to the learning curve, there should be a gradual increase over time. Increases greater than expected indicate that some changes have occurred that require immediate investigation and resolution. Productivity by operator graphed over numerous pay periods also needs to be evaluated and greater deviations than expected should also be investigated to insure that proper reporting procedures are being followed.

3. Does management understand and fully support the reasons and need for the revision?

Management must possess a detailed knowledge of the system and the fundamental foundation on which it is based and must comprehend the need for and understand the changes that justify the revisions. This is the purpose of the ECO system.

4. How involved should the employees be in the revision of the incentive rates?

Prior to the initiation of new studies. those employees whose rates may be modified must be advised that the new time studies will begin to examine the effect of the changes that have occurred in certain processes. These changes must be explained in detail to enable the employees to realize and acknowledge that the changes have actually occurred and ask questions if desired. Each employee has the right to ask any questions since this affects their earnings and employment status.

5. How is the incentive rate calculated?

See Glossary.

6. How is incentive pay calculated?

See Glossary.

For example: Assume the operator produces 30 units that earn 1.25 standard hours per unit in a given week during which he works 38 hours on incentive but has 2 hours he cannot work due to the unavailability of components.

The number of standard hours he earned is 30 units × 1.25 standard hours per unit or 37.50 standard hours earned in 38 hours. His productivity is (37.50 standard hours earned)/(38 hours worked) or 125% for those 38 hours. His pay would be calculated by multiplying his base rate per hour × 125% × 38 hours and adding two times his base rate for those hours he was not able to work due to the unavailability of components. The time he is idle and cannot work due to the unavailability of parts is referred to as downtime.

The percentage of downtime that an operator incurs may indicate that a problem exists somewhere in the system. The percentage should be close to zero, anything greater than zero could indicate an imbalance which warrants further study, the greater the deviation from zero, the greater the problem. The downtime reports create another situation for the supervisor; he has to insure their accuracy and total the hours and report them to payroll since the operator earns incentive pay on the hours worked less the downtime hours. Regular pay is earned on downtime hours.

7. Does the operator have to wait on parts before he can start working? If he must wait, then preceding operation must be studied to determine the cause of the imbalance. The time needed to complete the preceding process should be somewhat greater than the following one to prevent the operator from having to wait on a component to begin work.

8. Is the group to be studied represented by a union?

The only difference this will make is that after notifying the supervisor of the intent to take work measurement studies, the next person to notify is the person representing the union. Together if the representative desires, then the person selected for the study should be approached. This applies to each time a practitioner desires to take a time and motion study. The first person is the supervisor then the union representative. Should the person for some reason that you ask choose not to participate in the study, then thank him and ask another person. His reason for non-participation is confidential and personal.

D. EXAMPLES OF SUCCESSFUL AND UNSUCCESSFUL IMPROVEMENTS

1. EXAMPLES OF SUCCESSFUL IMPROVEMENTS

a. In the example discussed in the previous chapter beginning on page find the sewing department in a furniture plant, sewers were paid based on production standards that were no longer applicable. The standard rates did not differentiate between model numbers, the type of material sewn and the existence of welt cords, and the shape of the item being sewn,

After installation of the new system, the number of sewers was reduced to twelve, the average wages of the sewers actually increased. During the period the sales volume of the company increased. Thus the lack of documentation of the old incentive system prevented any attempt to address any obvious flows.

The studies must be evaluated periodically, at least annually, to determine if there have been any changes in materials or processes involved. This is the reason that detailed elemental descriptions are critical with the date of the study and are so important in the maintenance of labor standards since how can a change be justified if one cannot document that a change has occurred.

b. An example of method improvement in establishing direct standard hours occurred in a sewing department of a furniture plant. Due to changes that had occurred over time in management and the industrial engineering staff, an attempt to determine how the existing standards had been established was unsuccessful. An examination of existing standards indicated that improved direct labor standards need to be established for a number of reasons:

1. No documentation was discovered to provide details as to the derivation of the existing rates.
2. Rates were provided for small, medium, and large seats and backs with no explanation as to models were included in each size.
3. The rates for sewing fabric and vinyl were identical.
4. The rates for sewing with or without welt cord were the same.
5. The rates did not consider the shape of the item being sewn.

The critical questions that were asked prior to the initiation of studies.

1. How will the sewing operators be paid if there is no work available?

    If possible they should be transferred to another department if work is available for which they are trained. Otherwise they would be sent home.
2. How will an operator be compensated if asked to work in another department?

    Based on the author's experience they should be paid at the higher of the two rates.
3. For those operators whose earnings are less than the designated threshold, what remediation steps will be taken?

    Training should be made available until the operator attains the desired productivity level or is transferred to another department.
4. How will the work to be performed be distributed to insure equitable distribution?

    The work will be distributed as it becomes available. The supervisor will focus on assigning work as soon as it becomes available to insure that the department's schedule is met in a timely manner and to minimize the unforced idle time among operators. Since the rates were verified for accuracy, each operator should be capable of earning identical percentages.

The improvements that were initially made in the direct labor standards resulted in increased productivity in the sewing department, the transfer of some operators into other higher paying jobs in the plant, fewer defects, and a reduction in the standard direct labor costs for sewing seats and backs.

Another huge benefit that will increase over time is the documentation resulting from the formal time studies taken. The documentation enabled the development of a standard data sheet that could be used to accurately determine the direct labor standard hours and the cost of sewing new or proposed products.

2. AN EXAMPLE OF A FAILURE VERIFICATION OF INCENTIVE RATES

An example of an installation that failed occurred in a plant in which the author had the privilege of working as an outside employee to verify the accuracy of the incentive rates for an incentive system for its 400 plus plant employees that had been established by an outside industrial engineering firm. The company manufactured window coverings for various applications. An individual incentive system was established for each department over a period of several weeks. Anyone who did not meet the production standard within a specified period of time was subject to termination.

As an industrial engineer with many years' experience in the establishment of incentive rates in both union and nonunion plants, after several weeks of studying and analyzing the rates, the author realized that the rates were fair and reported that to management.

The implementation was a failure because the following critical questions were not asked:

1. Were existing methods evaluated to determine to what extent methods improvements could have been implemented prior to the initiation of the studies?

    This is the most critical aspect of any project. Tasks that are not needed must be eliminated. To quote Dr. Peter Drucker, "There is nothing so useless as doing efficiently that which should not be done at all." Methods improvement must occur before the initiation of any project especially if the project is the installation of an incentive system. The result of the failure will result in the earning of higher incentive pay due to the operators' realization that opportunities for improvements exist and initiating those improvements due to their desire to earn more with little or no additional work.

    The goal of the company was to install the incentive systems as soon as possible with the belief that savings in direct labor cost would be realized after installation. As a result, there was an insufficient effort to evaluate possible method improvement efforts. Operators did make improvements with no revisions in direct labor standards to reflect the improvements and consequently direct labor incentive pay increased. This is an example to illustrate that the perceived short-term gains would be more than offset by the cost increases in direct labor.

2. Should the company have advised the employees of the goal of installing an incentive system prior to the beginning of the studies. If so at what point should employees been notified?

    In the opinion of the author, the employees should have been notified desire of the company to install an incentive system and the reasons for its installation as soon as possible after the decision had been made. There are numerous reasons but the author's opinion is that no one appreciates surprises especially those that may affect a paycheck. The author when traveling used motel accommodations with a company that advertised that the best surprise was no surprise.

3. How much involvement in the studies should the employees have had?

The employees are the key to the success of the company and any program. The employees should have been fully engaged from the beginning of the project. Training on incentive systems should have been made available on company time for all employees since employees not in the system may at a later date be included in an incentive system.

The training needs to explain in detail the methodology of determining the rates, an interpretation of the meaning of the rates, the effect the rates will have on potential earnings, the goal the company hopes to achieve as a result, and the efforts the company will make to those who need training to meet the minimum levels of output needed to remain employed and earn incentive pay.

The plant engineer should have been involved throughout the process to enable him to address issues or questions as the studies are being conducted and after the studies have been completed.

4. How much involvement in the studies should the plant engineer have had to enable him to answer questions after the consulting company leaves?

The plant engineer should have been intricately involved in the establishment of all rates. After the consultants leave, it will be the responsibility of the plant engineer to maintain these systems which cannot be done without a clear understanding of how they were established and to answer any question the operator may have.

5. Why was there not a training program established in anticipation of meeting the needs of employees who were on the borderline of meeting production goals?

The result of the failure to ask and answer these questions was that the employees discussed the feasibility of unionization and invited union representatives to hold meetings for the purpose of forming a union. The problem was not the possibility of unionization; it was the distrust in management that resulted and the loss in time and communications between employees and management. The result was required costly management meetings with all supervisors with an out-of-town labor attorney concerning proper conversation with employees and resulted in delayed shipments and cost overruns. These management meetings were mandatory and resulted in a lack of supervision on the production floor.

This implementation did not have to end in failure. Tragically the plant ultimately closed resulting in the loss of over 400 jobs and the various tax-associated revenues. These failures were preventable. One huge factor working against it was the desire of management to implement this incentive system as soon as possible in order to reap the economic benefits. Management failed to think critically through the situation to determine the worst possible outcomes. In the end, this failure cost the employees, the vendors, the community, and the company.

As an industrial engineer who has installed several incentive systems, an incentive system is beneficial, in the author's opinion, only if it is fair to the employees first and then fair to the company. But before the company can begin installing a system, the employees must first be educated as to when and why and be constantly updated during the entire process. The author was involved in implementing an incentive system that involved 26 employees that required six months to implement, four to acquire the data, and two to validate the data.

Before implementing any standard for the record in the opinion of the author, it should be implemented on a trial basis with the results known only to engineering and management. The reason at that point is that the amount of unforced idle time is unknown. Before time studies were taken, studies were taken to improve the methods involved in the manufacturing of the product, employees will learn and initiate on their own new and improved techniques that did not materialize during the initial methods improvements studies.

An incentive system is one that rewards an employee based on quality output. In the manufacturing environment, an industrial engineer would establish a standard time for each routing operation. Standard time has been defined as the time required by the average skilled operator working at a normal pace to perform a specified task, using a specified method, allowing time for personal needs, fatigue, and delay (26). For an employee working alone, his pay would be computed by multiplying the number of products produced times the standard hours per product. This multiplication yields earned hours. The percentages included for fatigue and delay would vary with each individual job and must be evaluated taking all the tasks required for each job into consideration.

For an operator working on an assembly line, the controlling operation or the bottleneck on the line determines the production standard since that operation dictates the maximum production for that line. The line should be as balanced as possible to minimize unforced idle time among employees as tasks are being performed. Part of an industrial engineer's function is to equalize the work among all employees so that wait time between employees is minimized as well as work in process inventory.

Total pay is the base rate times the incentive rate percentage time the number of hours worked on incentive plus the number of hours worked at base rate or the number of hours worked while on downtime. The number of hours worked on incentive plus the number of hours worked at base rate must equal the total hours worked by the employee.

Incentive systems can be based on an individual or a group. Regardless of the method used, employees are paid at their base rate when they are unable to work due to a lack of work and which this period is referred to as downtime. This time must be recorded and subtracted from total hours to calculate the percentage of incentive pay earned. See Glossary for an explanation of the calculation of incentive rate and incentive pay earned. If a line is paid on a group incentive basis, the percentage of idle time should be constantly monitored to insure that work continues to flow smoothly along the assembly line.

Other types of group incentive pay include bonuses which are incentives that are paid to employees who exceed output goals. These incentives are in addition to their base pay rate. Commissions are another type of incentive pay. These are usually associated with sales positions as real estate or retail. Less common types are incentive pay for knowledge and skill that reward employees with higher pay as an incentive for the increased knowledge or skills acquired. Another type is profit sharing and stock option plans (PSSOP). Profit sharing is a scheme in which employers return a portion of net profit to their employees on compliance with certain service conditions and qualifications. The purpose of introducing profit-sharing schemes has been mainly to strengthen the loyalty of employees to the firm by offering them an annual bonus (over and above normal wages) provided they are on the service rolls of the firm for a definite period. The share of profit of the worker may be given in cash or in the form of shares in the company. These shares are called bonus shares (27).

6 Installing Measurement in a Recently Installed or Existing Individual Incentive System

A. BRIEF OVERVIEW OF THE IMPORTANCE OF MEASUREMENT IN AN INDIVIDUAL INCENTIVE SYSTEM

A properly installed individual incentive system provides opportunities to earn additional monies for employees in the system, improves morale, increases organizational productivity, and increases job satisfaction.

B. CRITICAL QUESTIONS TO ASK BEFORE INSTALLING MEASUREMENT IN A NEWLY ESTABLISHED INDIVIDUAL INCENTIVE SYSTEM WITH SUGGESTED SOLUTIONS

1. How is an individual operator output recorded?

    Based on the experience of the author, a real-time RFID system can track individual output and convey production by order to subsequent processing operations.

    The barcoding system will have the correct part number, part count, the standard hours earned for each completed part, and the operator number built into the code. This information in addition to the hours worked by each employee will be used to calculate the productivity of each employee. The number of hours worked by each employee will be extracted from the time clock data.

2. How are employees paid if there is no work available?

    This time is usually referred to as downtime. For an individual operator downtime should be close to zero. The operator should be at his workstation except during breaks and lunch performing assigned tasks. When no work is available, employees will need to maintain a daily log of the time they had to stop work and when they could restart due to the availability of parts and the length of time they were unable to work. This log must be verified by the supervisor. The number of hours that the operator cannot work is subtracted from the hours available to yield hours worked.

DOI: 10.1201/9781003153412-6

These are then divided into earned hours to determine the incentive rate percentage earned by that employee for that period. Total pay is calculated by multiplying the base rate times by the number of downtime hours plus the (total hours worked less downtime hours by the incentive percentage earned). The incentive percentage earned is the (number of units produced times the standard hours per unit)/the total hours worked less the downtime hours.

The number of downtime hours that each operator records is important to maintain. First, if the operator is scheduled to work an eight-hour shift and records two hours of downtime, then that means that 25% of the time he was not working at his assigned tasks.

This is a measure that must be followed by day of week and by operator. It could indicate that a problem exists prior to this operation that prevents components from getting to this operation on a timely basis? It could indicate that there is a need for training of operators and supervisors.

This underlines the importance of accurate documentation.

a.   The total amount of downtime. This needs to be analyzed by operator, day of week, and day of the month.

b.   The percentage efficiency earned. This needs to be analyzed by operator, day of week, and day of the month.

C.   CRITICAL QUESTIONS TO ASK FOR POSSIBLE POTENTIAL NEED TO REVISE MEASUREMENT TOOLS IN AN EXISTING INDIVIDUAL SYSTEM

1. Have any changes occurred in the manufacturing process since the direct labor standards were established?

   Examples of these changes include a change in materials used, tasks that need to be performed by an operator, a change in product quality specifications, the use of automation, a relocation of a work center, a change in the skill level needed by the operator, or a change in the workplace environment of working conditions.

   A comparison of the elemental descriptions written to document the existing processes will result in any changes that may have occurred.

   Any and all changes are the justification of the initiating new studies to determine the extent of the changes. Any and all changes regardless of the effect must be documented in detail in the elemental descriptions as well as other associated documentation. Changes must have been approved through the normal ECO process.

2. Have incentive earnings for one or several operators changed more than the mean increase?

   Studies to determine the cause(s) of this change must be thoroughly investigated and changes implemented as needed.

3. How will operators be paid if they are unable to work due to the unavailability of components?

    The number of standard hours he earned is 30 units × 1.25 standard hours per unit or 37.50 standard hours earned in 38 hours. His productivity is (37.50 standard hours earned)/(38 hours worked) or 125% for those 38 hours. His pay would be calculated by multiplying his base rate per hour × 125% × 38 hours and adding two times his base rate for those hours he was not able to work due to the unavailability of components.

4. How is the difference in operator skill levels taken into consideration?

    The skill level would be compensated for in the rating factor given to the operators by the practitioner. This is another reason to insure that studies are conducted by different practitioners on the same operators on different days and to validate results.

5. Does the operator have to wait on parts before he can start working? If he must wait, then the preceding operation must be studied to determine the cause of the imbalance. The time needed to complete the preceding process should be somewhat greater than the following one to prevent the operator from having to wait on a component to begin work.

    How will temporary periods when operators cannot work due to lack of materials be treated?

    This time, often referred to as downtime, is frequently maintained by each operator separately and verified by the supervisor for accuracy.

6. Is the group to be studied represented by a union?

    The only difference this will make is that after notifying the supervisor of the intent to take work measurement studies, the next person to notify is the person representing the union. Together if the representative desires, then the person selected for the study should be approached. This applies to each time a practitioner desires to take a time and motion study. The first person is the supervisor then the union representative. Should the person for some reason that you ask elects not to participate in the study, then thank him and ask another person. His reason for non-participation is confidential and personal.

    As the practitioner is taking the time study, evaluate each step for improvement and the potential for the elimination of waste. Too often a process begins that was started when the line or process was established is seldom evaluated for improvement and continues due to habit.

    Each task must be evaluated. There is no rule that the task must be continued as is or at all.

7. What are elemental descriptions?

    Elemental descriptions are a detailed explanation of the steps in a task that the operator performs.

    See Chapter 3 for a detailed explanation of an example and the purpose.

8. What is the date of the study on which the current direct labor standards are based?

The studies must be evaluated periodically, at least annually, to determine if there had been any changes in materials or processes involved. This is the reason that detailed elemental descriptions are critical with the date of the study and are so important in the maintenance of labor standards since how can a change be justified if one cannot document that a change has occurred.

D. EXAMPLES OF SUCCESSFUL AND UNSUCCESSFUL IMPLEMENTATION OF METHOD IMPROVEMENT IN ESTABLISHING DIRECT STANDARDS

1. EXAMPLES OF A SUCCESSFUL IMPLEMENTATION

a. In the example presented in Chapter 5, there were several ambiguities that had to be eliminated. The first was the definition of the unit being measured. The units on which operators were defined as small, medium, and large. Due to poor documentation, there was no information that differentiated among the three different sizes. There was no documentation of the type of fabric sewn, the inclusion of a border, and welt cord.

The first action is taken to define precisely a unit that would satisfy the current situation and be applicable for all future sewn covers. Time studies revealed there was no statistical difference in the time needed to sew the different materials, fabric and vinyl, the addition of a border or the inclusion of a welt cord using a folder. Due to potential changes in material sewn and the methodology of the sewing process, the definition of unit sewn was expanded to include the chair style, the type of fabric used, the addition of a border, and the existence of a welt cord. During the establishment of the labor standards, fabric or vinyl did not receive additional processing to add fire protection, the reduction of the spreading of germs, or to meet increased environmental concerns as Leadership in Energy and Environmental Design (LEED).

To satisfy the current situation and potential changes, it was determined that a different part number would be created for each chair style, each type of material, and the existence of a border and welt cord.

Currently the labor standard for sewing a chair style was the same regardless of the material used, and the existence of a border and welt cord. To prevent future confusion and allow for expansion for the sewing of additional types of material and additional processes, a unique part number was assigned to each different cover sewn. The use of meaningful numbers was discussed and after determining that chair styles would never change and that all covers from the sewing department would be sewn, the chair style, and the letter S was included in each part number. All the part numbers were alphanumeric and contained the same number of characters to prevent the possibility of a mistake occurring due to the omission of a character.

This application of measurement was successful due to critical thinking that asked the following questions.

1. How could the problems with the existing incentive system have been prevented?

   These and future problems were prevented by anticipating possible future processes and material change. Detailed documentation was included to eliminate present and potential misunderstandings and misinterpretations. Illustrations were included for clarification.

2. How can the confidence of the operators be assured in the accuracy of the rates?

   The operators were involved in studies at the beginning and during the studies. The operators elected a spokesperson who was aware of the methodology used and kept informed of the progress.

b. The author worked at a textile plant that manufactured fabric in several widths that became casts material for broken bones.

The author used time studies to study establish the direct labor standards for the winding operation. The fabric was cut into different widths lengthwise, three, four, and five inches, depending on the intended application, the smaller widths were intended to be wrapped around the smaller limbs. Each width was rolled into a three-meter length and packaged five rolls per carton for shipping.

The process consisted of the operator opening a carton with the width, obtaining a roll the width specified on the order, placing the roll onto a back stand, threading the beginning of the roll onto the machine that would unroll, and layering the fabric vertically into one-meter lengths until there were three wraps per bundle. The operator would tape the wraps together to insure the integrity of the bundle. The operator would place five bundles into the carton, tape the carton closed, and then apply the correct label. Each label contained the product number, the date of manufacture, and the operator number. Each width was assigned a different product number.

The unit of measure for the operator was a carton consisting of five bundles. This unit of measure was used since the pricing structure was based on a unit of five.

The result of the implementation was an increase in productivity of 21%. This was due to the cooperation between the operator, the engineer, and management identifying and eliminating unnecessary or non-value-added tasks that the operator at one time performed.

This implementation was successful due to the asking of the critical questions;

a. What are the non-value tasks being performed by this operator?
b. Would relocating the process result in an overall improvement?

   The machines weaving the fabric were on the first floor which was the basement due to the weight of the machines, the current process was on the third floor, and the shipping department which inventoried finished goods for shipment was on the second floor. As a result of a value stream map, this process was relocated to the second floor to facilitate shipping.

2. AN EXAMPLE OF A FAILURE VERIFICATION OF INCENTIVE RATES

Incentive rates were established by an outside consulting firm for a company with over 400 employees. The consultants failed due to time constraints to explain to all employees the procedures that were to be used. As a result, rates were determined but were unable to be communicated in a meaningful manner to the employees due to their lack of involvement. The plant engineer did not participate in the development so he was unable to answer any of the questions of employees and their supervisors.

This failure could have been avoided if the following critical questions had been asked and answered.

1. How can the employees better understand the methodology used in determining labor standards?

    Understanding and appreciation of the methodology can be increased by the inclusion of employees in the actual taking of time studies to enable them to understand the reasons for the detainment and the actual calculations involved.

2. Can employees use the mean times for each elemental to determine improvement strategies?

    The training should include a detailed explanation of elemental descriptions to enable each operator to compare his time for the performance with the standard to discern the difference and make adjustments as needed.

7 Integrating Methods Improvement in the Revision of an Existing Group or Installation of a New Existing Group Incentive System

A. BRIEF REASON FOR THE EXISTENCE OF A GROUP INCENTIVE SYSTEM

A group incentive system encourages teamwork and a sense of cooperation and responsibility and simplifies data collection.

B. THE FOLLOWING CRITICAL QUESTIONS MUST BE ASKED AND ANSWERED BEFORE INITIATING STUDIES TO IMPLEMENT A NEW GROUP INCENTIVE SYSTEM?

1. Is there a system in place to record production?

   This is the first action item that must be addressed. Production must be recorded in the routing file as units per hour. Based on the author's experience, the most accurate is automatic using bar codes that are read electronic. Depending on the use, the scanning equipment on the line will provide the most accurate and recent information.

   The use of bar codes will enable a change to be detected immediately whether positive or negative and provide feedback to management and the individual operator as to the progress toward goals.

2. How are the rates to be established?

   As discussed in Chapters 2 and 3, there are four possible methods. The one preferred by the author is a formal time study. See Glossary for detailed instructions.

3. How will waste or products that cannot be used for production be recorded?

   Only usable products must be counted and the count has to be verified for

DOI: 10.1201/9781003153412-7

accuracy to insure that shipment shortages do not occur and that payrolls are accurate.

4. How will temporary periods that operators cannot work due to lack of materials be treated?

This time, often referred to as downtime, is frequently maintained by each operator separately and verified by the supervisor for accuracy. Operators are normally paid at their normal base rate during downtime.

5. How will an operator be paid if he is transferred to another department due to lack of work in his normal department?

Based on the experience of the author, the operator is paid the higher of the two base rates, the one at the job to which he was transferred or his old one.

6. How supportive should management be of the program?

This program as well as all programs must have the full and undivided support of all levels of management. This just does not mean posters and banners but real employee engagement.

7. How involved should the employees be in the establishment of the program?

Each employee should understand the computation of a rate, its derivation, and its meaning. Each employee has the right to ask any questions since this effects their earnings and employment status.

The percentage of downtime that an operator incurs may indicate that a problem exists somewhere in the system. The percentage should be close to zero, anything greater than zero could indicate an imbalance which warrants further study, the greater the deviation from zero, the greater the problem. The downtime reports create another situation for the supervisor; he has to insure their accuracy and total the hours and report them to payroll since the operator earns incentive pay on the hours worked less the downtime hours. Regular pay is earned on downtime hours.

8. How is incentive pay calculated?

See Glossary.

For example: Assume the group produces 30 units that earn 1.25 standard hours per unit in a given week during which the group works a total of 38 hours on incentive but has 2 hours the group cannot work due to the unavailability of components.

The number of standard hours he earned is 30 units × 1.25 standard hours per unit or 37.50 standard hours earned in 38 hours. The productivity is (37.50 standard hours earned)/(38 hours worked) or 125% for those 38 hours. The pay for each member in the group would be calculated by multiplying his base rate per hour × 125% × 38 hours and adding two times his base rate for those hour she was not able to work due to the unavailability of components.

9. What is the skill level of the operators?

The skill level would be compensated for in the rating factor given to the operators by the practitioner. This is another reason to insure that studies are conducted by different practitioners on the same operators on different days and to validate results

10. Does the operator have to wait on components before he can start working?

If he must wait then the preceding operation must be studied to determine the cause of the imbalance. The time needed to complete the preceding process should be somewhat greater than the following one to prevent the operator from having to wait on a component to begin work. The operator should not complete his assigned tasks too quickly since that will result in the succeeding operator being forced to wait until he completes the unit on which he is currently working.

11. Is the group to be studied represented by a union?

The only difference this will make is that after notifying the supervisor of the intent to take work measurement studies, the next person to notify is the person representing the union. Together, if the representative desires, then the person selected for the study should be approached. This applies to each time a practitioner desires to take a time and motion study. The first person is the supervisor then the union representative. Should the person for some reason that you ask choose not to participate in the study, then thank him and ask another person. His reason for non-participation is confidential and personal.

As the practitioner is taking the time study, evaluate each step for improvement and the potential for the elimination of waste. Too often a process begins that was started due to necessity and is seldom evaluated for improvement and continues due to habit.

The most accurate measurements are made by a skilled practitioner using a stopwatch or similar measuring device. The individual tasks must be documented that includes the tools or fixtures that must be used in the procedure. See Chapter 3 for a detailed explanation and purpose for elemental descriptions.

12. What is the date of the study on which the current direct labor standards are based?

The studies must be evaluated periodically, at least annually, to determine if there have been any changes in materials or processes involved. This is the reason that detailed elemental descriptions are critical with the date of the study are so important in the maintenance of labor standards since this documentation will show a change.

13. What are the advantages of a group incentive system versus an individual one?

The group plan encourages teamwork to enable team members to increase their incentive earnings by members increasing their skill levels to enable them to perform additional tasks to reduce idle time of an individual operator. All members realize that idle time on the line results in less earnings since the employees are paid at their base rate during idle time.

14. There is less documentation to maintain. On an individual incentive system, the production and downtime of each operator must be recorded. For a group system, only the production of the group and the downtime for the group need to be documented. Team members will work together to reduce downtime to increase incentive earnings.

C. CRITICAL QUESTIONS TO ASK IF REVISING AN EXISTING GROUP INCENTIVE SYSTEM

1. Why is there a need to revise the existing group incentive system?

There are various reasons to revise an existing system. Changes in the manufacturing process as the addition of a robot to reduce material handling, the change of a vendor, a change in the product specifications, a redesign of workstations or packaging would require a revision to the existing system.

2. Another reason is to replace a poorly designed one or one that has not been maintained and updated as needed. If this is the primary reason, then the effected employees must be advised that the existing system is obsolete due to reasons that are not the fault of the employees, that their cooperation in the new studies is appreciated, and most importantly every effort will be made by the company to prevent this situation from reoccurring. During the studies and after installation of the new system, employee earnings should not be decreased because this is a failure of management and not the employees.

D. EXAMPLES OF BOTH SUCCESSFUL IMPLEMENTATION OF MEASUREMENT IN A GROUP INCENTIVE SYSTEM

AN EXAMPLE OF A SUCCESSFUL IMPLEMENTATION

A group incentive system can be employed in any situation that involves more than two employees that seek improvement. The author was involved in the establishment of successful group systems in the safety department. (see p. 161 for additional information).

AN EXAMPLE OF AN UNSUCCESSFUL IMPLEMENTATION

One plant in which the author worked as a consultant manufactured window coverings for sliding glass doors. The coverings were produced on an assembly line that consisted of eight workstations. An outside consulting company had installed an individual incentive system for each operator on the line.

Due to time constraints, the outside consulting company failed to investigate improvement opportunities and established rates on the current assignment of tasks for each operator. The assembly line was poorly balanced resulting in the occurrence of frequent bottlenecks during production. Again, due to time constraints there was no effort made to cross-train employees to enable them to assist other operators to prevent the development of the bottlenecks. The result was that there was significant unforced idle time for each operator since if there was no work available, the operator simply waited until work did become available. Employees who are unable to work at their assigned tasks due to the lack of work are paid at their base rate and must maintain records of this time. This record of downtime must be verified by management since it affects the amount of earnings for each employee. Each operator on the line recorded the beginning and ending for each time he was unable to work. These times were totaled for each employee and verified by the supervisor.

The author conducted a study via work sampling and discovered that the mean downtime for each operator was 40% of the assigned work day. This equated to over three hours each day that the operator was idle due to the failure to obtain a better-balanced assembly line. The author proposed that the individual incentive system be replaced with a group incentive to encourage operators to learn other tasks that would decrease the unforced idle time due to bottlenecks and provide an opportunity to earn incentive pay which was greater than the base rate.

The author was unsuccessful in this attempt to install a group incentive system.

8 Integrating Measurement in the Revision of an Existing Group Incentive System or Installation of a New Existing Group Incentive System

A. BRIEF OVERVIEW OF THE NEED FOR MEASUREMENT IN A GROUP INCENTIVE SYSTEM

Measurement enables a comparison of outcome over time to insure that improvements are achieved. Critical to measurement is the definition of what is being measured and consistency in its use over time.

B. CRITICAL QUESTIONS THAT MUST BE ASKED BEFORE INSTALLING A NEW GROUP INCENTIVE SYSTEM WITH SUGGESTED SOLUTIONS

1. What are the advantages of a group system over an individual incentive system?
   a. It encourages each team member to acquire additional skills to enable each team member to reposition himself on the production line as needed to insure a continuous flow of product and to minimize the accumulation of work-in-process.
   b. It reduces the amount of production information that needs to be collected since output only needs to be recorded at the end of the line.
   c. It simplifies the calculation of incentive pay since all team members receive the same percentage.

2. Is there a constraint on the production line?
   The constraint must first be minimized before studies can be initiated to install the group incentive system. This methodology is fully explained in Chapter 3 [46].

DOI: 10.1201/9781003153412-8

3. Is there a system in place to record production?

This is one of the first action items that must be addressed. Production must be recorded be as in the routing file as units per hour. Based on the author's experience, the most accurate is electronic using barcodes. Depending on the use, scanning equipment on the line will provide the most accurate and recent information.

The use of barcodes will enable a change to be detected immediately whether positive or negative and provide feedback to management and to the individual operator as to the progress to goals. The barcoding system will have the correct part number, part count, the standard hours earned for each completed part, and the operator number built into the code. This information will be recorded and uploaded.

4. After the line has been balanced then time studies need to be taken only on the operation that is the constraint since this is a limiting factor on the output. Each operator on the production line should be cross-trained to perform several tasks to enable each operator to prevent an accumulation of work-in-process inventory.

5. How will waste or products that cannot be used for production be recorded?

Only usable products must be counted and the count has to be verified for accuracy to insure that shipment shortages do not occur and that payrolls are accurate.

6. How will temporary periods that operators cannot work due to lack of materials be treated?

This time, often referred to as downtime, needs only be maintained by the constraining operation since the remaining processes can continue. Operators are normally paid at their normal base rate during downtime. Based on the experience of the operator, operators should be cross-trained to minimize downtime to enable continuous processing of the product.

7. How will an operator be paid if he is transferred to another department due to lack of work in his normal department?

Based on the experience of the author, he is paid the higher of the two base rates, the one at the job to which he was transferred or his old one.

How supportive should management be of the program?

This program as well as all programs must have the full and undivided support of all levels of management. This just does not mean posters and banners but real employee engagement.

8. How involved should the employees be in the establishment of the program?

Each employee should understand the computation of a rate, its derivation, and meaning. Each employee has the right to ask any questions since this affects their earnings and employment status,

a. There are a number of effective incentive plans for groups which are designed to promote teamwork. Several existing plans include: [47]

1. The Pressman's plan.

   This plan takes into consideration the productivity of all the employees in a group. Pay for one period is based on the proportion of the standard that the group achieved in the current period compared to the previous period. This plan encourages operators to prevent the avoidance of unforced idle time by learning new skills needed to perform the tasks of the preceding and succeeding operator.

2. Scanlon plan

   A bonus paid to employees when their performance is or productivity is increased against a given standard by a certain percentage. Variables over which the employees and the company have some control are emphasized. These plans usually contain suggestion programs or idea programs most of which have monetary rewards.

9. How will incorrect part quantities be determined and what effect will this have on incentive pay?

   The shortage or overage will appear at a late stage of assembly. The shortage or overage will be investigated as to the cause to determine a solution. The incentive pay of the employee will be changed accordingly.

10. How are employees paid if there is no work available?

   This time is usually referred to as downtime. For an individual operator, downtime should be close to zero. The operator should be at his workstation except during breaks and lunch performing assigned tasks. When no work is available, employees will need to maintain a daily log of the time they had to stop work and when they could restart due to the availability of parts and the length of time they were unable to work. This log must be verified by the supervisor. The number of hours that the operator cannot work is subtracted from the hours available to yield hours worked. These are then divided into earned hours to determine the incentive rate percentage earned by that employee for that period. Total pay is calculated by multiplying the base rate times by the number of downtime hours plus the (total hours worked less downtime hours by the incentive percentage earned). The incentive percentage earned is the (number of units produced times the standard hours per unit)/the total hours worked less the downtime hours.

The number of downtime hours that each operator records is important to maintain. First, if the operator is scheduled to work an eight-hour shift and records two hours of downtime, then that means that 25 percent of the time he was not working at his assigned tasks.

This is a measure that must be followed by day of week and by operator. It could indicate that a problem exists prior this operation that prevents components from getting to this operation on a timely basis? It could indicate that there is a need for training of operators and supervisors?

This underlines the importance of accurate documentation,

a. The total amount of downtime. This needs to be analyzed by operator, day of week, and day of the month.
b. The percentage efficiency earned. This needs to be analyzed by operator, day of week, and day of the month.

C. CRITICAL QUESTIONS TO ASK BEFORE ATTEMPTING TO INSTALL MEASUREMENT IN AN EXISTING GROUP INCENTIVE PLAN

1. Have any changes occurred in the bottleneck process since the direct labor standards were established?

   Examples of these changes include a change in materials used, tasks that need to be performed by an operator, a change in product quality specifications, the use of automation, a relocation of a work center, a change in the skill level needed by the operator, or a change in the workplace environment of working conditions.

   A comparison of the elemental descriptions written to document the existing processes with existing processes will result in any changes that may have occurred.

   Any and all changes are justification for initiating new studies to determine the extent of the changes. Any and all changes regardless of the effect must be documented in detail in the elemental descriptions as well as other associated documentation. Changes must have been approved through the normal ECO process.

2. Have incentive earnings for the group changed more than the mean increase?

   Studies to determine the cause(s) of this change must be thoroughly investigated and changes implemented as needed.

3. How will operators be paid if they are unable to work due to the unavailability of components?

   They will be paid at their base rate.

4. How is incentive pay calculated?

   See Glossary.

5. How is the difference in operator skill levels be taken into consideration?

   The skill level would be compensated for in the rating factor given to the operators by the practitioner. This is another reason to insure that studies are conducted by different practitioners on the same operators on different days and to validate results.

6. Does the operator have to wait for parts before he can start working? If he must wait, then the preceding operation must be studied to determine the cause of the imbalance. The time needed to complete the preceding process should be somewhat greater than the following one to prevent the operator from having to wait on a component to begin work.

How will temporary periods that operators cannot work due to lack of materials be treated?

This time, often referred to as downtime, is frequently maintained by the operator on the bottleneck operation verified by the supervisor for accuracy.

7. Is the group to be studied represented by a union?

The only difference this will make is that after notifying the supervisor of the intent to take work measurement studies, the next person to notify is the person representing the union. Together if the representative desires, then the person selected for the study should be approached. This applies to each time a practitioner desires to take a time and motion study. The first person is the supervisor then the union representative. Should the person for some reason that you ask elects not to participate in the study, then thank him and ask another person. His reason for non-participation is confidential and personal.

As the practitioner is taking the time study, evaluate each step for improvement and the potential for the elimination of waste. Too often a process begins that was started when the line or process was established and is seldom evaluated for improvement and continues due to habit.

Each task must be evaluated. There is no rule that the task must be continued as is or at all.

8. What is the date of the study on which the current direct labor standards are based?

The studies must be evaluated periodically, at least annually, to determine if there had been any changes in materials or processes involved. This is the reason that detailed elemental descriptions are critical with the date of the study and so important in the maintenance of labor standards since how can a change be justified if one cannot document that a change has occurred.

D. EXAMPLES OF SUCCESSFUL AND UNSUCCESSFUL IMPLEMENTATIONS

1. AN EXAMPLE OF A SUCCESSFUL IMPLEMENTATION OF A GROUP INCENTIVE SYSTEM

The author had the privilege of managing successful safety programs with two different companies. Success was defined as a continued reduction in the number of accidents and injuries at the various plant locations. Although successful, the author sought a methodology to achieve further reductions.

Both companies had multiple plants with an excellent safety culture supported by all levels of management. Both companies published a newsletter that publicized events that occurred at each plant and when possible recognized individual employees for their various achievements. The author as a result of the frequency with which he visited each plant realized that not only did individual plants appreciate recognition but individual employees particularly enjoyed seeing their name and achievement in the newsletter distributed quarterly.

The author began including the safety team members in each newsletter and the particular reductions in the injury and accident rates that the safety team had achieved. Also included were the actions of each team to achieve the results as a learning tool for other plants due to the similarity of many processes among the manufacturing facilities.

During routine visits to other plants with each company in his role as industrial engineer, the author would be approached by employees in management and others employed at that plant with the suggestion of some type of competition among the plants to reward the plant that reduced the accident and injuries the most.

The author proposed to management a safety incentive program to reward the plant with the greatest percent reduction in accident and incident rates from the previous quarter by providing the employees of the winning plant the opportunity to receive a 50-dollar gift card via random drawing. The initial drawing was for 50 employees. All employees in the winning plant were eligible since anyone could correct an unsafe act.

The savings in the cost of workman's compensation more than paid for the gift cards. Additional benefits were an improvement in attendance, a reduction in employee turnover, and an overall increase in morale.

2.   AN EXAMPLE OF AN ATTEMPTED GROUP INCENTIVE SYSTEM

One plant in which the author worked as a consultant manufactured window coverings for sliding glass doors. The coverings were produced on an assembly line that consisted of eight workstations. An outside consulting company had installed an individual incentive system for each operator on the line.

Due to time constraints, the outside consulting company failed to investigate improvement opportunities and established rates on the current assignment of tasks for each operator. The assembly line was poorly balanced resulting in the occurrence of frequent bottlenecks during production. Again due to time constraints there was no effort made to cross-train employees to enable them to assist other operators to prevent the development of the bottlenecks. The result was that there was significant unforced idle time for each operator since if there was no work available, the operator simply waited until work did become available. Employees who are unable to work at their assigned tasks due to the lack of work are paid at their base rate and must maintain records of the time. This record of downtime must be verified by management since it affects the amount of earnings for each employee. Each operator on the line recorded the beginning and ending for each time he was unable to work. These times were totaled for each employee and verified by supervision.

The author conducted a study via work sampling and discovered that the mean downtime for each operator was 40% of this assigned workday. This equated to over three hours each day that the operator was idle due to the failure to obtain a better balanced assembly line. The author proposed that the individual incentive system be replaced with a group incentive to encourage operators to learn other tasks that would decrease the unforced idle time due to bottlenecks and provide an opportunity to earn incentive pay which was greater than the base rate.

The author was unsuccessful in this attempt to install a group incentive system.

9 Integrating Methods Improvements in a New or Existing Manufacturing Process Change

A. BRIEF OVERVIEW OF A MANUFACTURING PROCESS AND THE NEED FOR CONTINUOUS EVALUATION TO SEEK OPPORTUNITIES FOR IMPROVEMENT

A manufacturing process exists only to change raw materials into a usable finished product. Functions as transportation, handling or storage are not a part of the manufacturing process because they do not participate directly into the conversion of raw materials into a finished product. Rapidly developing technology and consumer needs will require a continual evaluation of current processes for an improvement of any processes to enable an organization to thrive and remain competitive. These same drivers of change will necessitate a continual review of existing processes for potential cost-effective improvements.

B. CRITICAL QUESTIONS TO ASK BEFORE INITIATING A NEW MANUFACTURING PROCESS WITH PROPOSED SOLUTIONS

1. Why is there a need for a new manufacturing process?
    There are several reasons to initiate a new manufacturing process:
    A. To introduce a new product to the modified product to the market place
        For a new product, what is the potential sales volume in units?
        What is the material and direct labor costs for the new product?
        a. The bill of material (BOM) with needed indentations must be documented with specifications and conversion factors for each level. The costs and vendors of purchased items must be documented with quotes. The quotes must specify free on board (FOB), origin or FOB destination. FOB origin means that the customer is responsible for the freight and insurance costs of getting the product to him whereas FOB origin means that the manufacturer incurs the freight and insurance costs of delivery.

DOI: 10.1201/9781003153412-9

    b. Process routings must be specified for each manufactured compo-
nent to insure manufacturability. The routings must include any
fixtures or software needed for CNC (computer numerically controlled) equipment. The use of existing standard data sheets or the
use of existing routings will assist in this task. The costs for any
items that need to be purchased need to be determined to enable the
amortization per unit for these costs.

    The author based on experience recommends that a product be
produced to insure that the item can be manufactured to meet or
exceed specifications before that item is made available to be purchased profitably.

    c. Overhead costs as material handling and utility expenses must be
anticipated and included as warranted.

2. To introduce a modified product to the market place

    The bills of materials and the process routings of the initial product
must be examined in detail to determine the effects of the modified
product. These effects must be examined in detail to insure that any
changes in the purchased items in the bill of specifications are available
and that any cost of modification in tooling, fixtures, and software are
considered.

    The projected sales volume and the cost of production will be
required to compute profit levels and determine the feasibility of the
project.

    The process can be an addition, a replacement for an existing one, a
combination, or a deletion.

    Each option must be studied to determine the net effects of this
change. The net effects must be considered before a final decision is
rendered.

C. CRITICAL QUESTIONS TO ASK BEFORE REQUESTING TO INITIATE A CHANGE IN AN EXISTING MANUFACTURING PROCESS WITH PROPOSED SOLUTIONS

1. Why is this request being submitted?

    There are numerous causes for the need to change a manufacturing
process.

    A quality specification may have changed that requires an additional
task to be added or deleted. For example, an item at one time that required
a smooth surface no longer needs a smooth surface.

    A vendor may have changed the unit of measure supplied that could
result in an additional task that must be added or one that may be deleted.
As an example, a vendor may begin supplying material in sheets that require
one or more cutting operations that previously provided the material to the
correct dimensions.

A process may be automated that previously was manual requiring a modification in the layout of the operation. An existing machine used in the process is to be upgraded or removed.

A new product or variations of an existing one may be added to the product line. For example, a furniture manufacturer may offer a new finish on existing items.

2. Who can request a change in a manufacturing process?

The primary functions in requesting a change in existing processes are marketing, purchasing, quality assurance, and engineering.

3. What reasons exist for the functions stated above to initiate a request?

Marketing would request a change to modify an existing product, to add a product, or to discontinue manufacturing a product due to ever-changing market and customer needs and expectations.

Purchasing would request a change if a current vendor could no longer supply an existing material with current specifications, a vendor was able to supply a material with improved cost or delivery or terms, or supply a material with improved quality and manufacturability.

Engineering would request a change to reflect a new application of technology, methods, processes, regulations, an addition or elimination of operation to support product additions, modifications, or deletions.

Quality assurance would initiate a request to modify specifications of products manufactured or any materials used in the manufacturing process.

4. What is the process to initiate and insure the correct implementation of these requested changes?

The function requesting the change must initiate an engineering change order (ECO). Different originations would require different changes in documentation.

If marketing is the initiator, then the different types of changes requested by marketing would affect different documentation. If the request was to modify an existing product, then engineering would need to evaluate the affect in the bill of material, process routings, process instructions, and drawings, and note the changes required in the ECO. The material planner would establish dates on which the changes were to occur. Repair and service parts must be included to determine the effects and note needed document modifications on the ECO. If the request was to add a new product, then engineering would be required to determine manufacturability; provide necessary drawing; determine the materials, fixtures, and processes that could manufacture; package the new item to meet or exceed specifications; and note the information in the ECO. Purchasing would be to acquire vendors if new materials were needed and include the revisions needed in the ECO. If the request is to discontinue a product, then engineering must determine changes needed in bills of materials, process routings, process instructions, and fixtures used in the manufacturing of that product and note needed changes in the ECO. The material planner must consider the effect that discontinuing production not manufacturing will have on the unique materials used in that product.

Purchasing would request a change if a current vendor was no longer supplying an existing material with current specifications; a vendor was able to supply a material with improved costs, delivery, or terms; or supply a material with improved quality and manufacturability. Quality assurance would need to evaluate the proposed material to insure that it meets or exceeds existing specifications. Engineering would need to determine the documentation affected by the change in vendors and provide that information on the ECO. The material planner would establish the effectivity dates the changes would occur based on existing inventory and the delivery date of the new material.

Engineering would request a change to reflect a new application of technology, methods, an addition or elimination of operation to support product modifications, additions, or deletions. If engineering is the initiator, then engineering must determine the documentation affected and the effects the change will require and note this information on the ECO.

Quality assurance would initiate a change to modify existing specifications on manufactured products. Engineering would determine the documentation affected and the changes needed and note that on the ECO. The material planner would determine the effectivity date based on existing inventory and note the date on the ECO.

After approval of the ECO by all effected functions, the ECO is then given to the responsible function to make the needed changes as indicated in the ECO and verifies the accuracy of the changes.

1. If a machine is to be acquired for a new process or to replace an existing one, these questions must be asked to insure that long-term usage costs are minimized.
   a. Will the machine be purchased or leased?
       This decision will be based on numerous factors which include:
       1. The expected lifetime of the machine.
       2. The availability of cash, or the interest rate on borrowed money if a loan is considered.
       3. The method of depreciation and the amounts of the expenses taken, financial condition of the company.
       4. The length and the depth of the warranty associated with the machine.
       5. The rate at which technology changes in the industry. The greater the velocity of technology change, the shorter the new machine will be involved in the manufacturing of a product with rapid obsolesce.
       6. The anticipated maintenance cost of this machine initially and its long-term costs. The location of the machines manufacturer must be considered if technicians from the manufacturer are needed to augment the skills of local maintenance personnel.
       7. The skills needed by those who will operate the new machine. If the skill level must be elevated, then the time and costs of those skill upgrades need to factored into the timing and installation of the new machine.

2. What other companies are using this machine?

Discuss with engineering and maintenance personnel of other companies to obtain objective accurate information concerning the performance of this machine. This information should include other machines considered prior to the acquisition, reasons for the decision to acquire the current one, maintenance records, utility consumption, ease of repair and maintenance, and the productivity and efficiency of the machine.

Functions that must be included in the decision include purchasing, finance, engineering, and maintenance. Other functions that may need to be included are human relations, safety, and materials planning.

D. EXAMPLES OF SUCCESSFUL AND UNSUCCESSFUL MANUFACTURING PROCESS CHANGES

1. SUCCESSFUL EXAMPLES OF PROCESS CHANGES

a. The author observed a potential opportunity in the manufacturing of paper tubes that had a latex application. Latex is a type of aqueous adhesive that is nontoxic and odorless. It has numerous applications as in the manufacture of furniture, now-woven fabrics, paper processing, and wireless binding. Beneficial characteristics include excellent initial tackiness, a long shelf life of one year, and no harmful materials in composition [48].

The change involved the installation of an electric eye that would detect the beginning point of the paper tube as it was extruded from the mandrel. A spray gun would spray a controlled amount of latex onto the tube at the beginning of the extrusion process and stop when the tube has completely left the mandrel. This change not only reduced the cost of the latex but eliminated the need to elevate the five-gallon bucket onto the stand four feet high and the inserting of a slot ¼ inch by 1 inch into the bottom and insured that correct alignment of the latex onto the tube. Also, the amount of latex that did not flow onto tubes flowed but onto the floor into the waiting drain that fed into the local sewer system was greatly reduced.

This discussion resulted in effecting a successful manufacturing change in a process involved in the application of latex adhesive on the outside of a paper tube. The latex was purchased in five-gallon buckets and applied to the outside of paper tubes to provide an adhesive surface to grab a material to initiate the wrapping process. It was applied by having the bucket mounted in a metal stand over the tube as it was removed from the machine by the drilling of a slot in the bottom of the bucket that provided a continuous flow of the latex. Tubes were produced continually but since the length of the tubes varied, there was a gap between tubes that varied as the length of tubes. Since the flow of the latex was continuous, some of the latex flowed onto the tubes but some was wasted and flowed into the drain and ultimately the local water supply.

The manufacturing process consists of installing a roll of paper board into a stand, feeding the roll through a section of rollers that applies adhesive and maintains a constant pressure, cuts the paperboard to the specified length by the activation of a knife as the paperboard triggers an electric eye which is located above the mandrel. A carriage transports the length into a slit in a mandrel that becomes the diameter of the tube, is rolled into a tube as the mandrel is rotated in a counter clockwise direction, and the recently wound tune is removed from the mandrel by an arm attached to the end of the carriage onto a conveyor belt that conveys the tubes into an oven to dry the adhesive.

The production rate as measured by tubes per minute was limited by the carriage and did not vary with the length of the tube but did vary inversely with the width of the paperboard used to manufacture the tube due to having less control with a wider sheet.

The production standard was not based on the length of the tube. Frequency of roll changes varied inversely with the tube length. A production standard did exist for roll changes which varied directly with the width of the roll installed on a pneumatic stand.

The industrial engineering technician who calculated productivity daily was also given the additional task of determining the total number of linear inches of tubes produced each day, making sure to include the waste attributable to the latex application broken down by the thickness of the tube to ascertain waste varied by thickness. The amount of waste generated during manufacturing and the number of defects was decomposed by length and the percentage did not vary with the length of the tube.

The critical questions that were asked that facilitated this into becoming a successful application were

1. What other methods exist for applying latex?

   The team investigate three methods, namely brushing, application with a roller or similar device, and spraying. Due to the curvature of the surface, the need for consistency in the amount applied, and the ability to control the amount applied by order, the option of using a pneumatic spray gun was chosen due to the fact that it was the easiest one to install and maintain.

2. Will additional manpower be needed?

   The existing crew consisted of three people, namely an operator, an assistant and a bundler. The bundler assisted during setups and roll changes as did the assistant operator. Tasks associated with changeovers included relocating an electric eye that determines the length of the tube. The new method involved two additional electric eyes, one placed at the beginning of the disposal of the tube from the mandrel and the other at the discharge end. The relocation of these two electric eyes would require no more than one minute and could be easily added to the existing tasks of the existing crew.

3. What will be installation cost for this new method?

   Conversion to this new process requires the purchase and installation of an additional electric eye and spray equipment that could easily pump latex from the bucket that was purchased into the spray gun. As the tube was removed from the mandrel, the electric eye activated the spray gun and deactivated it as the tube left the mandrel.

Inventories of pneumatic hoses were maintained due to normal deterioration as were air guns. Allocation of maintenance hours to projects or machines was not kept, so the determination of the cost of installation cost was not possible in this situation. An estimate of $2000 was used, which was known to be excessive. As a rule of thumb, if unsure of estimating, always estimate costs to be greater than expected since it is always better to have the project to realize a savings greater than expected rather than less than expected. Always use exact cost from quotes if available.

4. What are the savings for this new method?

Savings will result from the difference in usage:

The difference in usage between the methods

a. This difference must be multiplied by the total number of linear inches of tube produced during that period.

b. The savings for that period is the savings from the step above multiplied by the cost per pound of latex, FOB plant.

c. Analysis of the results revealed that the waste percentage between the two methods of application did not vary as the application varied, thus the percentage was constant regardless of the method of application.

d. Sales for convolute tubes are maintained in increments of thousands (M).

Thus the savings for the period must be annualized and the installation cost be deducted.

e. Result is annual savings, reciprocal is payback.

f. What effect will changing the method of latex application have on the process of manufacturing the paper tubes?

The question was asked to crew members and management and the unanimous answer was that there would be no effect. This was confirmed by the author with random sampling studies and relatedly constant productivity and efficiency percentages.

g. What is the revenue earned from the latex?

The industrial engineering technician calculated the revenue from the upcharge for several months after the change. The revenue earned was compared with the cost of latex consumed during the same period and this project was very cost effective.

5. It is crucial after the conclusion of a project to compare the savings gained or additional revenue earned to the cost of the project for several reasons;

a. To validate the costs and savings as depicted in the project proposal.

b. To learn and gain experience to enhance project management.

6. How will this affect purchasing?

The purchasing agent stated that after the project was completed the frequency of purchases would be reduced.

7. Will additional training be required for the machine operator?

The maintenance department discussed in detail the operation of the spray system with all members of the crew and demonstrated the air pressure required, and fully trained all members of the crew on each shift in its operation.

b. If a customer ordered a carpet grade tube with a length of 50 inches, the machine would produce a tube that was 152 inches in length, allowing for waste and bundle them with 19 to a bundle and transport them to the re-cutting department and place them into an empty bundling rack (Figures 9.1 and 9.2).

This department consisted of one re-cutter operated by one operator. This operator cut the bundle with steel band cutters. The operator would pick up one tube from the rack and slide the tube to be cut onto the mandrel, which was 36 inches high to the center of the mandrel, cut one inch off one end by depressing the foot pedal and disposing of the waste into an adjacent carton to make it smooth, move the remainder of the tube to the designated stop position (preset at 50 inches), depress a foot pedal which activated the cutter to cut the tube at the required length, grasp the cut tube and place into a bundling rack, return to the re-cutter, and continue this process until the bundling rack contained 19 tubes. At this time the operator would strap the bundle with three steel straps and wait for a fork lift to remove the bundled tubes so

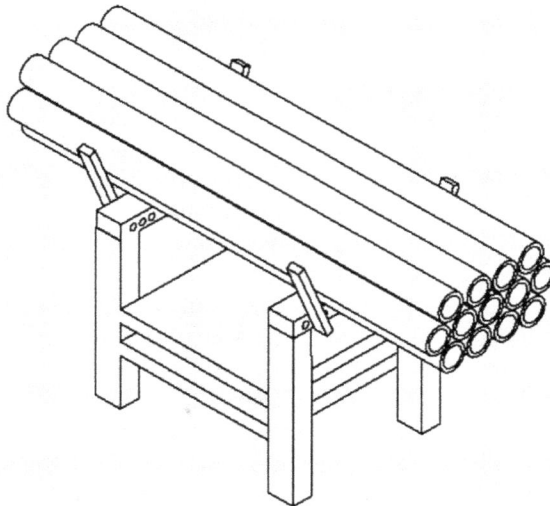

FIGURE 9.1   Fixture for racking tubes manually that can be adjusted for various outside diameters of tubes (orthogonal view).

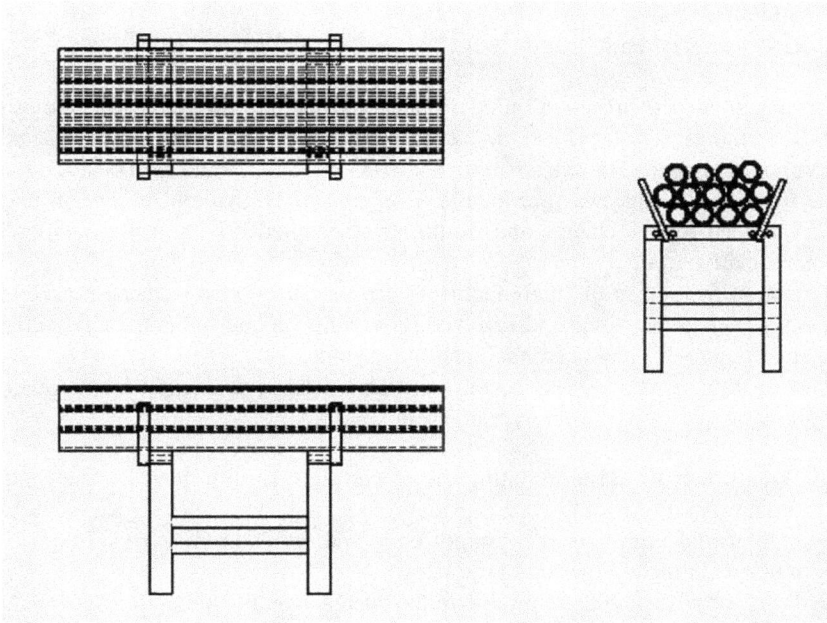

FIGURE 9.2   Fixture for racking tubes manually that can be adjusted for various outside diameters of tubes (plan, top, and right hand side views).

that the process can begin. This process continued until all the tubes with the original length had been recut.

Although there was a re-cutting upcharge for this operation, the author realized that the extra operation fell into the category of overprocessing. He began discussing possibilities with management and maintenance for cutting the tubes to length during processing. Both suggested a knife that could cut the paperboard as it was being rolled into a tube. However, several questions would have to be addressed.

1. Currently after re-cutting both ends of the tune were smooth as if they have been sanded.
2. Currently after re-cutting the tubes are cut to the precise length with tolerances within ⅛ inch of the required length.

Cost savings were compared to the price upcharge and since savings exceeded the additional costs, the company realized that it was advantageous for customers to accept these modifications.

Management agreed that price reductions should be offered since the customer was accepting less perceived quality in the product. These perceived product losses were:

1. Smooth ends.
2. Greater variation in length.

The engineer and the salesman called on each customer that would be affected by the proposed change. The engineer first explained the current process and discussed the proposed change. The salesman then stated that there would be a cost reduction if the proposed change was acceptable. Every customer without exception understood the need for the change and would have accepted it without the savings but was appreciative of it.

This modification to the manufacturing process succeeded because it asked the necessary critical questions and involved those who needed to be, namely management, maintenance, plant personnel, and the customer.

This example was a success because critical thinking was employed to generate the following questions which were answered.

1. What effect would eliminating the re-cutting operation have on the final product?

    It would eliminate the smooth ends and result in length variation as much as ½ inch in the final product.
2. Were these variations allowable in the current specifications?

    The current specifications did not mention smooth ends or length tolerances but these tubes had been delivered to customers over many years. Compared to the current product, the customer may have thought the new product was of lesser quality which is the reason a discount was offered.
3. Did the change produce cost savings?
    a. Direct labor cost decreased from the elimination of the separate re-cutting operation.
    b. Direct labor costs increased from the two additional bundlers needed when these tubes were being produced.
    c. The net result was a savings in direct labor costs.
    d. Indirect labor costs were also decreased due to not having to deliver tubes to the re-cutter department and then from the re-cutter department to shipping.
    e. Throughput was also increased since the tubes did not have to undergo a separate time-consuming operation.
4. Payback equals cost installation cost divided by annual direct labor cost savings.

a. Costs that need to be determined:

1. Costs of re-cutting per M tubes designated as C(R).
2. Costs of direct labor for the two additional people needed to bundle tubes with the new process- designated as C(B).
3. Cost of two additional bands of steel strap (½ inch × 0.020) used to encapsulate tubes designated as C(D).

4. Sales discount given to accept tubes with less "perceived" quality designated S(D).
5. Savings per M could be calculated as C(R) less the sum of C(B), C(S), and C(D).

A percentage for waste must be added since each tube that was to be recut was manufactured with two additional inches to accommodate waste, one inch on either end; the shorted tube manufactured for re-cutting was 100 inches in length. Using this length will provide the maximum percentage which will yield minimum savings.

b. Net savings equals savings less waste

Annualized savings can be computed by calculating the ratio of savings during this time period and annualizing the data. The reciprocal is payback.

This company did not have a standard cost system in effect at the time this change was implemented; therefore, it was not possible to determine incremental revenue from the re-cutting upcharge to the incremental costs to corroborate the net savings.

Vladimir Lenin, "doveryai, no proveryai," which translated means "trust, but verify," made famous by President Ronald Reagan. This phrase was repeated many times by President Reagan during various meetings with the Russian leader Mikhail Gorbachev (35).

2.   EXAMPLES OF UNSUCCESSFUL PROCESS CHANGES

a. One furniture company as employer of the author had a case goods plant and one that manufactured seating in the same town. The sales volume of the case goods plant had increased such that the company felt that the volume could support a second plant. The case goods and the seating plant operated independently; the only asset shared was the computer system and the warehouse.

Although the variety of products included desks of different shapes and sizes, bookcases with different widths and heights, lateral files of varying heights and widths to credenzas with varying widths, there was a sufficient number of each of the main types of products, desks, bookcases, lateral files, and credenzas of the common sizes so that justified the relocation of these products into an existing building approximately 100 miles distant. The primary plant would continue to produce all the component parts needed and send these to the new facility for assembly, finishing, inspection, and packaging. The primary reason for the change was that direct labor in the new facility was 40% less than in the existing one.

The rationale for the division of manufacturing was that the existing plant had the equipment, the skilled plant personnel, engineering, maintenance, and the computer system that documented the current processes in detail. Management believed that the assembly operation could be easily taught and be learned with a few instructions. The finishing crew in the existing plant had many years' experience and produced an excellent product. The finishing crew in the new plant was new and due to their inexperience and lack of training and a trained supervisor, according to records maintained by the QA technician, produced scrap at a rate approaching 50%. This level of scrap resulted in the primary plant having to produce additional parts often resulting in overtime. The assembly plant also had to absorb extra hours due to the extra scrap. Over a six-month period, this scrap level was eventually reduced to 40% resulting in additional hours at both the original and the assembly plant.

Over time a number of changes in the manufacturing processes of case goods had occurred at the original plant. These changes may have involved the bill of materials, the routing, a vendor, a drawing, a fixture, or process routing but due to shortage of engineering personnel, the appropriate documentation was never submitted and the computer system was never updated. Another possible reason for a change is that these were highly skilled individuals who initiated a better method of achieving a task and implemented it of their own accord.

With several years both plants ceased operations. Failure was due to the lack of asking the following questions at a minimum:

1. Why is there such a difference in the cost of labor that initiated the plan to transfer the manufacturing of case goods to the distant plant?

   The difference was due to skill and experience level, the location of the current plant which contained highly skilled employees with many years' experience; the plant to which product was being transferred had a few employees, with some experience in assembly, few in basic manufacturing, and none in finishing.

2. How much training will be needed and what will that training cost be?
   Due to the close proximity, a two-hour drive, of the existing plant, the belief that existing personnel could provide training as needed at the new site.

3. Are the computer system, bills of material, routings, drawings, vendor files, and other needed documentation correct?
   The old system had never been revised or updated. Since this system had the Y2K problem as well, management decided to move manufacturing and modify the computer system rationalizing that, after all, manufacturing processes had been made to improve the accuracy of routings, bills of materials, inventory, and suppliers.

4. What is the real purpose of this change?
   If the goal was to save money, this proposed change in a manufacturing method did not achieve this goal. Within several years not only was the original case goods plant shut down, but the plant to which the product was eventually transferred was shut down as well causing the loss of over 600 jobs, loss of tax revenues and other losses. In the opinion of the author, this loss and many others like this did not have to occur.

b. Another example of a failure that occurred during change in a manufacturing process was owing to failure to ask the critical question determining if the plant was fully able to manufacture a product before it was illustrated in a catalog with prices so that orders could be taken. Because the company had been manufacturing desktops for many years, it assumed that it possessed the knowledge, skill, and manufacturing capability of producing this particular tabletop to meet the quality and quantity specifications required. The product was a new design that involved a metal insert that was to be placed into the tabletop.

The insert was to be placed in a groove cut into an elliptical shape into the desktop, requiring the precision that is achievable with a CNC router or similar machine. A CNC machine can have as many as five axes and is a motorized maneuverable tool controlled by a computer programmed to perform certain operations which are programmed into the software.

The situation was that the company had a CNC machine that was not capable of accepting updated software due to its age. Thus though the company did have a CNC machine, it did not have software that would enable the programming of adding a curve that would connect two straight sections thus completing an ellipse. No effort was made to produce a sample before the catalog was printed with prices to insure manufacturability of the item.

Attempts were made to produce a fixture that would allow manufacturing to route this curve manually but after numerous efforts with several different operators, management recognized that further efforts were futile. Routing of the top occurred after the veneer had been cut and edged glued into a face which is several inches larger in each dimension to accommodate further processing; this face is then glued onto a piece of medium density fiberboard (MDF) on one side and a sheet of paperboard on the bottom, this assembly is then edge-banded and the final operation is the routing operation. Thus if a desktop is scrapped at this step in the manufacturing operation, it has a large percentage of its material and direct labor costs (Figures 9.3 and 9.4).

The first experience of the author with a furniture company was with a small one located in rural NC. It produced a variety of low-end dinette sets consisting of tables, chairs, and barstools. The chairs and barstools could be purchased in a number of different combinations of fabrics and vinyl. The tables were produced in a variety of styles in several laminates. Prototype products of new styles were actually built to insure that the products could be actually produced and to eliminate any production problems before they were shown in the semi-annual Furniture Market in High Point, NC. For every product shown, an accurate bill of material and routing was determined prior to being shown and entered into a computer program that determined the actual cost. For any product, anyone in the company was able to visualize on the computer or print the BOM, the actual cost, the vendor, the purchased costs, the routing of the product, the sales history, the current inventory, and demand history. For those products that were shown at the market, anyone in the company was able to visualize or print everything except the inventory and demand history since

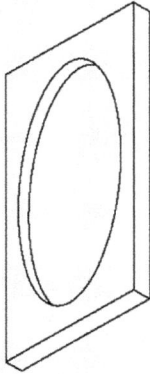

FIGURE 9.3   Desktop for the insertion of an inlay – (orthogonal view).

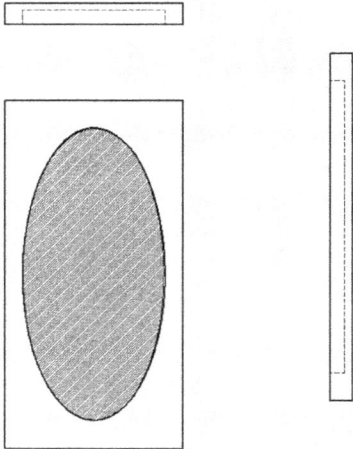

FIGURE 9.4   Desktop for the insertion of an inlay – (plan, top, and right hand side views).

none existed. Before any product was introduced at the market, the standard direct labor and material cost, the processes involved in producing the product, the vendors, order quantity, delivery terms were known, since if an order from a vendor at the semi-annual market were to be received, the company must be able to deliver the product profitably when desired and in the quantity required. The vendor may order any quantity ranging from one to a truckload with a delivery date between one and several weeks.

For an individual who had worked for a much smaller company that knew all vendors, the BOM, the routing, and the costs of a product before it was marketed, this situation was inconceivable for any company much less one of this size. This company was much larger when measured in sales, number of employees, number of plants, etc., but the product was advertised and priced without the assured knowledge of manufacturability. This would have never occurred at the smaller company for which the author had worked. The author is not sure how the larger company handled this situation, but he was sure that he would not want to admit to a large customer that the company sold a product that it had not produced a prototype of before marketing it.

The company had to withdraw the product from its catalog.

The critical questions that should have been asked and answered are:

1. Why was this not made and approved by sales and manufacturing before it was shown and priced in the catalog?

    The engineering department produces a sample to insure that the product can be produced. Industrial engineers use standard data sheets to determine the standard labor hours for production. Manufacturing engineering fabricates the necessary fixtures or obtains the needed software to insure that the components can be produced in the quantity desired to meet or exceed the quality standards. Packaging engineering insures that the current packaging will provide the necessary protection for the entire product line or modify or design additional packaging material as needed.

2. Why were not all the details of manufacturing specified first during the prototype phase of the project? The only reason that the author could surmise was the lack of management commitment to provide adequate resources. Without top management financial support, the project cannot succeed.

3. How can prices be determined before one is assured of manufacturability? The determination of prices must be based on the costs to manufacture the product. The calculation of costs includes the cost of the bill of materials, the vendors and the prices that will be paid to those vendors for raw materials, and the labor costs, both direct and indirect, for the manufacturing costs of all the processes.

    c. An example of a failed change in the manufacturing process occurred as the result of a furniture plant's desire to upgrade their cutting department from a manual operation to one using a CNC fabric cutter. Justifications for the machine would come from a reduction in direct labor, waste of fabric, a decrease of work in process inventory, and an increase in manufacturing throughput. The machine would replace six manual hand cutters.

The current process was to have the cutter roll out layers of identical fabric onto a table after the correct fabric was pulled for him from inventory, approximately 40 feet long, with as many layers as a vertical cutter would cut simultaneously, then locate the correct cardboard pattern hanging on the wall adjacent to the rolls of fabric, place the pattern pieces onto the fabric and adjust manually until the yield is maximized, clamp the fabric onto the table to prevent the fabric from moving once he has started cutting, and cut, mark, and set aside each piece into a small container that after completion is identified as the end product.

An evaluation was made of existing available machines on the market. Due to the potential savings it was realized that the sooner the equipment was purchased and installed, the sooner savings would accrue. The machine was to be used in a plant that was located in a small rural town in the middle of nowhere. The closest airport was at least a two hours' drive depending on traffic. It was decided to purchase a machine that was produced overseas due to the price difference of approximately 25K between the one made overseas and one produced in the United States.

The machine was received but did not meet US electrical requirements as specified in the purchase order but met the requirements of the country in which the fabric cutter was manufactured. Also the technical manuals for the cutter were in the language of the country in which the cutter was manufactured, which in this case was Spanish.

After proper electrical installation was made, then the machine began to cut fabric to everyone's expectations. The new process was that first all the pattern pieces for an item had to be digitalized or transferred into the computer and filed for easy retrieval. Then the CNC operator would roll out as many layers onto the bed of the machine as possible and then activate the machine. A vacuum held the fabric in place to prevent the fabric from moving after the machine began cutting the fabric. Cutting was quicker than manual as was ejection of the cut parts.

The problems with the machine occurred when the maintenance people in the plant were unable to make the necessary repairs since the technical manuals were in Spanish. Telephone call to the manufacturer was essentially useless due to language differences. Maintenance personnel from the company who produced the cutter were flown over to make repairs. Maintenance was not covered under warranty and the maintenance personnel from the producing company required first-class accommodations, vehicles, and meals. Eventually the maintenance personnel left and returned home. As a result of the inability of the maintenance department to make repairs as needed, the maintenance crew from the manufacturer became almost regulars but extremely expensive visitors. When the machine was not cutting fabric, the plant had to resort to the previous costly practice of manually cutting fabric.

    d. Another example of a failed change in a manufacturing process involved a company that manufactured windows and doors, producing vertical window/door coverings used to primarily cover sliding glass doors. The company purchased plastic material in eight foot lengths into which various items were inserted such as fabric, metal, etc. Slots for hanging above the sliding glass doors were then punched into these inserts to enable drawstrings to be attached for closure of the coverings. The median length of this material sold was 84 inches with a standard deviation of exactly one inch. The raw material was priced and purchased per linear foot. The process consisted of receiving the raw material in the warehouse, retrieving it by one person and bringing it to the first station in the assembly line which was a cutting table, cutting the eight foot material to the specified length and discarding the waste (mean of 12 inches into an adjacent trash container and paying for waste disposal), punching one end to accept the plastic insert and drawstring, moving it to the next workstation that attached the plastic inserts into the vanes and drawstrings, placing it onto a conveyor to enable assemblies to be added at the next workstation inside the headstock that enabled the vanes to rotate vertically 180 degrees, returning it to the conveyor as needed to enable quality assurance (QA) to perform an inspection

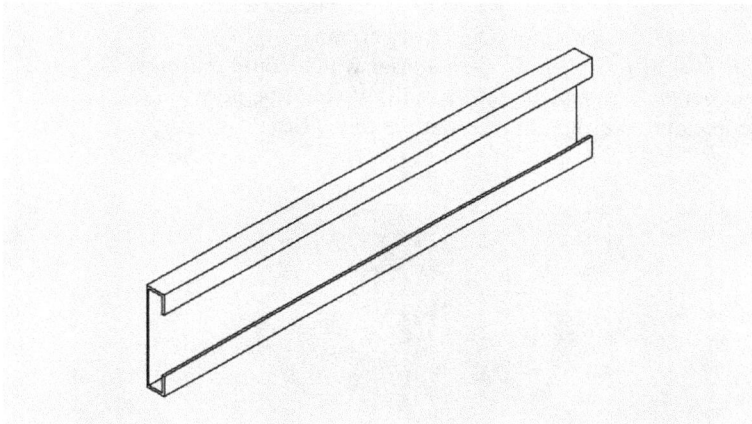

FIGURE 9.5   Vertical vane material for a fabric or meta insertion (orthogonal view).

FIGURE 9.6   Vertical vane material for a fabric or meta insertion – (plan, top, and right hand views).

to guarantee that the vane quantity was correct and the mechanism inside the head assembly functioned, and packaging the product at the next station for shipment.

The amount purchased of vane material exceeded 100K linear feet per year. The purchasing agent advised the author that was he was in the process of converting from purchasing 8 linear feet to 12 linear feet lengths to save the company an extraordinary amount of money.

The author analyzed the situation logically and realized that if this change was to be made, then the purchase of the additional four feet would result in the company losing money on this product due to two reasons:

1. Cost of waste increase by a factor of five because the waste would increase from one foot per vane to five feet per vane.
2. The cost of retrieval of raw material which could currently be handled by one person is doubled since handling would be now required to have two people due to its length (Figures 9.5 and 9.6).

10 Integrating Measurement into Changing an Existing or Developing a New Manufacturing Process Change

A. BRIEF OVERVIEW OF THE NEED FOR MEASUREMENT AFTER A CHANGE HAS OCCURRED IN A MANUFACTURING PROCESS

The purpose of the manufacturing change is improvement which could occur in numerous areas as an increase in the number of units produced per hour, a reduction in the number of defects produced per period of time, an improvement in the quality of the final product, a decrease in delivery time or an improvement in the working conditions, as a reduction in the noise level of a manufacturing process.

The units of measurement used before the change is implemented should be identical to those after the implementation, as man-hours per unit before the change and man-hours per unit after the change.

It is also crucial that the same measuring system, metric versus the US system be used since man-hours per ton is not the same as man-hours per pound is not the same as man-hours per kilogram. Confusion between the two is responsible for the loss of a 125-million-dollar satellite in 1998 [49] and errors in the administration of medications [50].

B. CRITICAL QUESTIONS THAT MUST BE ASKED BEFORE INTEGRATING MEASUREMENTS INTO THE DEVELOPMENT OF A NEW PROCESS CHANGE

1. Why is the process being added?

   The change could be the result of a change in the product specifications, unit of measure, a change in the packaging of the incoming material or finished product.

2. Who will initiate the ECO and asure that all effected documentation is changed correctly.

   The purchasing agent will initiate the ECO if there is a change in the unit of measure. This may affect the procedures for incoming raw material, the

DOI: 10.1201/9781003153412-10

cost, the inventory levels, the process routings and bills of materials, drawing, inspection instructions, etc. The engineer is responsible to insure that all affected functions are included and the accuracy of the changes. Quality would be responsible for a change in specifications, etc.

2. How will the manufacturing cost be affected.

All additional costs for the new or modified process must be determined and compared with the current costs. These costs will include direct and indirect labor, utilities and material changes. The initiator of the ECO with the assistance of the engineer will itemize and verify these costs.

3. Is the new process necessary?

The engineer must study the tasks of this new process be added to determine if these new tasks can be added onto the tasks of existing operators. If the tasks of existing operators and modified, then studies must be conducted to determine the effects on process routings, process sheets, inspection procedures, drawings, etc.

4. Does this new process require changes in material handling?

Studies must be made to determine what modifications to existing workstations need to accommodate these material handling changes.

5. Will this new process involve the use of hazardous materials?

Training, PPE, the methods for storage and handling of these materials must be evaluated to insure OSHA compliance and the risk of exposure is minimized.

C. CRITICAL QUESTIONS THAT MUST BE ASKED BEFORE INTEGRATING MEASUREMENTS INTO THE CHANGING AN EXISTING OF A NEW PROCESS

1. Why is the process being changed?

The change could be the result of a change in the product specifications, unit of measure, a change in the packaging of the incoming material or finished product.

2. Who will initiate the ECO and assure that all effected documentation is changed correctly?

The purchasing agent will initiate the ECO if there is a change in the unit of measure. This may affect the procedures for incoming raw material, the cost, the inventory levels, the process routings and bills of materials, drawing, inspection instructions, etc. The engineer is responsible to insure that all affected functions are included and the accuracy of the changes. Quality would be responsible for a change in specifications, etc.

2. How will the manufacturing cost be affected?

All additional costs for the new or modified process must be determined and compared with the current costs. These costs will include direct and indirect labor, utilities and material changes. The initiator of the ECO with the assistance of the engineer will itemize and verify these costs.

3. Is the change in the process needed?

The engineer must study evaluate the tasks in the existing process and compare them to the tasks after the process is changed to determine direct labor costs differences. This must be done to insure that is a cost-effective change. Included in the study will be the documentation effected and the modifications to the affected documentation.

4. Does this altered process require changes in material handling?

Studies must be made to determine what modifications to existing workstations need to accommodate these material handling changes.

5. Will this altered process involve the use of hazardous materials?

Training, PPE, the methods for storage and handling of these materials must be evaluated to insure OSHA compliance and the risk of exposure is minimized.

D. EXAMPLES OF SUCCESSFUL AND UNSUCCESSFUL APPLICATIONS OF MEASUREMENT TO MANUFACTURING PROCESS CHANGES

1. Examples of successful changes.
    a. The change involved converting from a continuous to a controlled spray application.

Information that is required to determine if this change is cost effective:

1. The cost of installation.

Conversion to this new process required the purchase and installation of an additional electric eye and spray equipment that could easily pump latex from the bucket in which it was purchased. As the tube was removed from the mandrel the electric eye activated the spray gun and deactivated it as the tube left the mandrel.

Inventories of pneumatic hoses were maintained due to normal deterioration as were air guns. Allocation of maintenance hours to projects or machines was not kept so the determination of the cost of installation cost was not possible in this situation. An estimate of $2,000 was used which was known to be excessive. As a rule of thumb, if unsure or estimating, always estimate high since it is always better to have the project to realize a savings greater than expected than less than expected. Always use exact cost from quotes if available.

2. Additional information was needed to evaluate the process change that required measurements
    a. The effect of this change on the direct labor costs for this process.

The existing crew of three people, an operator, an assistant, and a bundler. The bundler assisted during setups and roll changes as did the assistant operator. Tasks associated with changeovers included relocating an electric eye that determines the length of the tube. The new method involved two additional electric eyes, one placed at the beginning of the disposal of the tube from the mandrel at the discharge end. The relocation of these two electric eyes would require no more than one minute and could be easily added to the existing tasks of the existing crew.

This change had no effect on the direct labor costs which was confirmed by the author with random sampling studies and relatively constant productivity and efficiency percentages.

b. The effect of this change on the costs of the paperboard and the adhesive used in the manufacture of these paper tubes.

Studies were taken after the process change to determine the amount of waste and defective products were produced. The results of the studies conducted before the process change were compared to those taken after the change with the result that there was no change in the number of defective products.

c. The reduction in the amount of latex applied to shippable product.

Prior to the initiation of the project an industrial technician noted the installation of a full five-gallon bucket on the operator's production record. As this five-gallon bucket and its replacement were consumed, the industrial technician maintained records on the number of tubes produced with latex applied and the length of each tube. This enabled the calculation of the number of gallons required per M-linear inch of quality tubes produced.

The industrial technician then maintained the same information after the conversion to the new method. Tubes prices per M tube were based on length and as was the upcharge for latex application.

To calculate savings due to the process change, the difference between the two methods of application in gallons of latex per M-linear inch must be compared. The savings per M-linear inch is the difference in usage of latex in gallons per M-linear inch times the current cost of latex per gallon delivered to the plant.

The resulting calculations revealed that this was a very cost-effective project.

3. Another successful example of a manufacturing change is the elimination of a process.

If a customer ordered a tube for rolling carpet that could be cut to the desired length as the tube was being produced, then the need for recutting the tube was eliminated. This also improved throughput since it eliminated the travel to the recutting department and to the trailer that would be delivered to the customer.

The figure below was originally presented in Chapter 9 (Figures 10.1 and 10.2).

This department consisted of one recutter operated by one operator. This operator cut the bundle with steel band cutters, three bands per bundle. The operator would pick up one tube from the rack and slide the tube to be cut onto the mandrel of the recutter, which was 36 inches high to the center of the mandrel, cut one inch off one end by depressing the foot pedal and disposing of the waste into an adjacent carton to make it smooth, move the remainder of the tube to the designated stop position (for

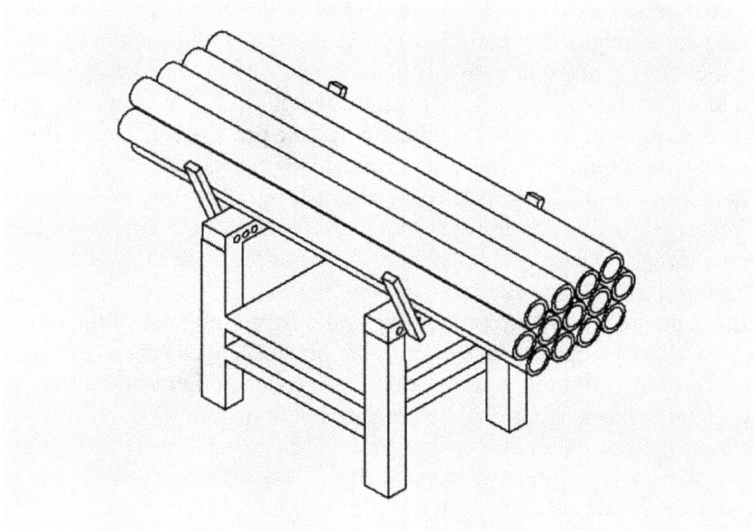

FIGURE 10.1   Fixture for racking tubes manually that can be adjusted for various outside diameters of tubes (orthogonal view).

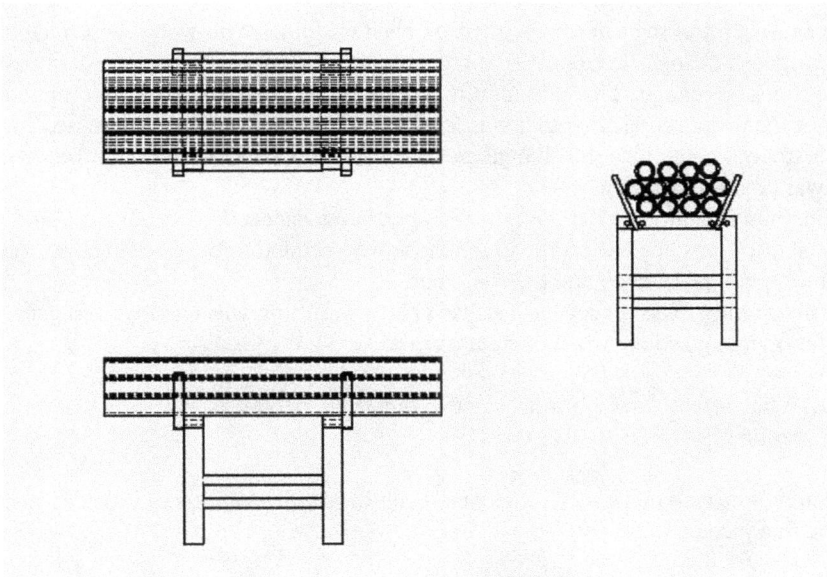

FIGURE 10.2   Fixture for racking tubes manually that can be adjusted for various outside diameters of tubes (top, plan, and right-hand side views).

this customer preset at 50 inches), depress a foot pedal which activated the cutter to cut the tube at the required length, grasp the cut tube and place into a bundling rack, return to the recutter and continue this process until the bundling rack contained 19 tubes. At this time the operator would strap the bundle with three steel straps and wait for a forklift to remove the bundled tubes so that the process can begin. This process is continued until all the tubes with the original length had been recut.

The author realized that the extra operation fell into the category of overprocessing. He began discussing possibilities with management and maintenance for cutting the tubes to length during processing. Both suggested a knife that could cut the paperboard as it was being rolled into a tube.

The tubes manufactured after recutting would have both ends of the tube smooth as if they have been sanded and are cut to the precise length with tolerances within ⅛ of an inch of the required length. The tubes produced that were not recut would not have smooth ends and a greater tolerance of half an inch.

Cost savings were compared to the price upcharge and since savings exceeded the additional costs, the company realized that it was advantageous for customers to accept these modifications.

Management agreed that price reductions should be offered since the customer was accepting less perceived quality of the product. These perceived product quality losses were:

1. Smooth ends
2. Greater variation in length

The engineer and the salesman called on each customer who would be affected by the proposed change. The engineer first explained the current process and discussed the proposed change. The salesman then stated that there would be a cost reduction of the proposed change that was acceptable. Every customer without exception understood the need for the change and would have accepted it without the savings but was appreciative of it.

This modification to the manufacturing process succeeded because it asked the critical questions needed and involved those who needed to be, management, maintenance, plant personnel, and the customer.

This example was a success because critical thinking was employed to generate the following questions which were answered.

1. What effect does eliminating the recutting operation have on the final product?

It would eliminate the smooth ends and result in length variation as much as ½ inch in the final product.

2. Were these variations allowable in the current specifications?

The current specifications did not mention smooth ends or length tolerances but these tubes had been delivered to customers over many years. Compared to the current product, the customer may have thought the new product was of lesser quality which is the reason a discount was offered.

3. Did the change produce cost savings?

Direct labor cost decreased from the elimination of the separate recutting operation.
Direct labor costs increased from additional bundlers needed when these tubes were being produced.
The net result was a savings in direct labor costs.
Indirect labor costs were also decreased due to not having to deliver tubes to the recutter department and then from the recutter department to shipping.
Throughput was also increased since the tubes did not have to undergo a separate operation.

Payback equals cost installation cost divided by annual direct labor cost savings.

4. Information that was required to determine the feasibility of the project
   a. The existing direct labor standards for this recutting operation was established with formal time studies and used to calculate the direct labor costs. The standard crew size was one operator.
   b. The one additional bundler needed increased the crew size from three to four. Studies were taken with the new method of the tubes. The result of these studies revealed that the production standards would not be affected. The standard direct labor costs would be increased due to the additional bundler.
   c. Measurements of the amount strap used for the various thicknesses of tubes that were recut were taken.
      Sales discount given to accept tubes with less "perceived" quality designated S(D)
   d. A sales discount of 5% was agreed upon by management. This agreed upon was not the result of any measurements but an estimate sufficient to encourage customers to accept the modified product. Net savings are expected to greatly exceed the discount.
   e. A cost of the waste generated of waste must be added since each tube that was to be recut was manufactured with two additional inches to accommodate waste, one inch on either end, the shorted tube manufactured for re-cutting was 100 inches in length. The amount of waste produced due to re-cutting operation was obtained through measurement. A waste study of the most recent 20 tubes was used to calculate a waste percentage. This percentage was used to calculate the savings from not producing the two additional inches of tube.

Net savings for this process change are computed by deducting the costs increases due to the additional bundler from the savings resulting in the elimination of the recutting operation and the two additional inches of tube. The cost discount provided to customers would reduce the net savings.

This company did not have a standard cost system in effect at the time this change was implemented; therefore it was not possible to determine incremental revenue from the recutting upcharge to the incremental costs to corroborate the net savings.

Whenever possible, savings must be verified by another method. To quote a favorite proverb of Vladimir Lenin, "doveryai, no proveryai," which translated means "trust, but verify" made famous by President Ronald Reagan. This phrase was repeated many times by President Reagan during various meetings with the Russian leader Mikhail Gorbachev. [51].

a. An example of an unsuccessful change

1. An example of a failed change in the manufacturing process occurred as the result of the desire of a furniture plant that wanted to upgrade their cutting department from a manual operation to one using a computer numerically controlled (CNC) fabric cutter. Justifications for the machine would come from a reduction in direct labor, waste of fabric, a decrease of work-in-process inventory, and an increase in manufacturing throughput. The machine would replace six manual hand cutters.

The current process was to have the cutter roll out layers of identical fabric onto a table after the correct fabric was pulled for him from inventory, approximately 40 feet long, with as many layers as a vertical cutter would cut simultaneously, then locate the correct cardboard pattern hanging on the wall adjacent to the rolls of fabric, place the pattern pieces onto the fabric and adjust manually until the yield is maximized, clamp the fabric onto the table to prevent the fabric from moving once he has started cutting, cut, mark, and set aside each piece into a small container that after completion is identified as to the end product.

An evaluation was made of existing available machines on the market. Due to the potential savings it was realized that the sooner the equipment was purchased and installed, the sooner savings would accrue. The machine was to be used in a plant that was located in a small rural town. The closest airport was at least a two-hour drive depending on traffic. It was decided to purchase a machine that was produced overseas due to the price difference of approximately 25K between the one made overseas and one produced in the United States.

The machine was received but did not meet US electrical requirements as specified in the purchase order but met the requirements of the country in which the fabric cutter was manufactured. Also, the technical manuals for the cutter were in the language of the country in which the cutter was manufactured, which in this case is Spanish.

After the proper electrical installation was made, the machine began to cut fabric to everyone's expectations. The new process was that first that all the pattern pieces for an item had to be digitalized or transferred into the computer and filed for easy retrieval. Then the CNC operator would roll out as many layers onto the bed of the machine as possible and then active the machine. A vacuum held the fabric in place to prevent the fabric from moving after the machine began cutting the fabric.

The problems with the machine occurred due to the inability of maintenance people to make the necessary repairs since the technical manuals were in Spanish. Telephone calls to the manufacturer were essentially unproductive due to language differences. Maintenance personnel from the company who produced the cutter were flown over to make repairs. Maintenance was not covered under warranty and the maintenance personnel from the producing company required first-class accommodations, vehicles, and meals. Eventually, the maintenance personnel left and returned home. As a result of the inability of the maintenance department to make repairs as needed, the maintenance crew from the manufacturer became almost regular but extremely expensive visitors. When the machine was not cutting fabric, the plant had to resort to the previous costly practice of manually cutting fabric.

2. Another example of a failed change in a manufacturing process involved a company that manufactured windows and doors that produced vertical window/door coverings used to primarily cover sliding glass doors. The company purchased plastic material in eight-foot lengths into which various items were inserted as fabric, metal, etc. Slots for hanging above the sliding glass doors were then punched into these inserts to enable drawstrings to be attached for closure of the coverings. The median length of this material sold was 84 inches with a standard deviation of exactly 1 inch. The raw material was priced and purchased per linear foot. The process consisted of receiving the raw material in the warehouse, retrieving it by one person and bringing it to the first station in the assembly line which was a cutting table, cutting the eight-foot material to the specified length, and discarding the waste (mean of 12 inches into an adjacent trash container and paying for waste disposal), punching one end to accept the plastic insert and drawstring, moving it to the next workstation that attached the plastic inserts into the vanes and drawstrings, placing it onto a conveyor to enable assemblies to be added at the next work station inside the headstock that enabled the vanes to rotate vertically 180 degrees, returning it the conveyor as needed to enable quality assurance (QA) to perform an inspection to guarantee that the vane quantity was correct and the mechanism inside the head assembly functioned, and packaging the product at the next station for shipment.

The amount purchased of vane material exceeded 100,000 linear feet per year. The purchasing agent advised the author that was he was in the process of converting from purchasing 8 linear feet to 12 linear feet lengths to save the company an extraordinary amount of money.

The author analyzed the situation logically and realized that if this change were to be made then the purchase of the additional four feet would result in the company losing money on this product due to two reasons:

a. Cost of waste increase by a factor of five because the waste would increase from one foot per vane to five feet per vane.
b. The retrieval of raw material which could currently be handled by one person is doubled since handling would now require to have two people due to its length Figures 10.3 and 10.4.

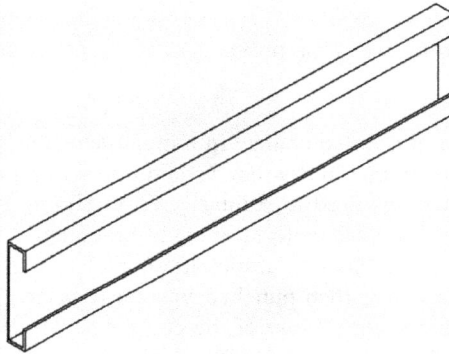

FIGURE 10.3   Vertical vane material for a fabric or metal insertion (orthogonal view).

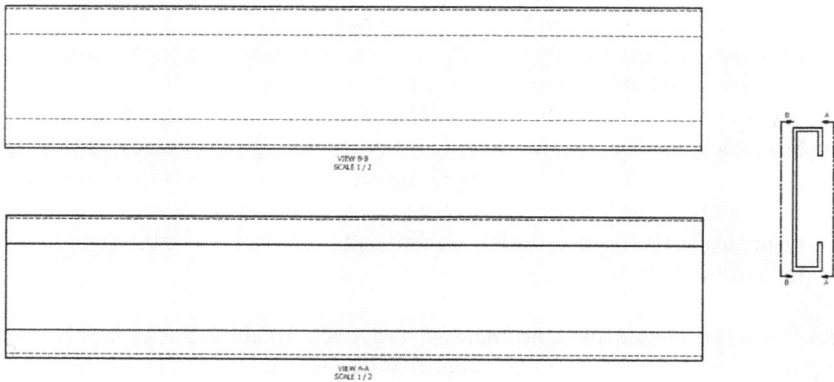

FIGURE 10.4   Vertical vane material for a fabric or metal insertion (top, plan, and right side view).

11 Integrating Methods Improvement in the Selection of a New Vendor or the Decision to Change an Existing Vendor

A. BRIEF OVERVIEW FOR THE NEED TO ADD A NEW VENDOR OR TO CHANGE AN EXISTING ONE

New products may be introduced when existing vendors are not able to supply one or more of the needed components of the new product. Specifications may change to the extent that existing vendors cannot supply materials that meet or exceed the new specifications. An existing vendor may cease operations or may decide not to continue to produce a particular material.

A vendor has a vital role in the success of any organization. As a result, there must exist an excellent and cooperative relationship between the two. Motivating factors for the selection of a new vendor differ and each motivating factor results in revisions to the various documentation that is essential to enable the organization to manufacture and deliver the product that exceeds or meets quality specifications. This documentation includes a minimum an item master file, a bill of materials, routing files, and a purchasing file.

The selection of a new vendor must consider numerous factors, including product quality, ratings from professional organizations, service history for existing customers, ISO certifications, prices quoted and terms of delivery.

B. CRITICAL QUESTIONS TO ASK PRIOR TO INITIATING THE SEARCH FOR A NEW VENDOR

1. What are the motivating factors that would cause the need for a new vendor?
    a. To obtain a better price, delivery or terms.
    b. To add a new component.
    c. To replace a vendor who no longer exists or can longer supply a needed item.

DOI: 10.1201/9781003153412-11

2. What is the methodology for initiating the search for a new vendor?

The function seeking a new vendor initiates an engineering change order (ECO). The initiator investigates if other functions will be affected by this change and includes the changes in the ECO. For example, the current vendor may supply the material in a roll which required a cutting operation to a certain size. The new vendor supplies the item to the specified size necessitating changes in the bill of material and the process routings for all the manufactured products that use this item. This ECO is distributed to each area affected by the change for approval. The material planner will establish an effectivity date for the changes.

3. What is the recommended method for selecting a new vendor?

The author recommends the development of a matrix that lists various numerous factors important to those functions affected and have each function rate its importance to the function on a five-point Likert scale. The results can then be tabulated and an optimal decision made.

The method of new vendor selection must:

1. Request the input of purchasing, manufacturing, receiving, accounting, and other functions as needed.
2. Revisited existing vendors periodically for needed changes.
3. Consider the initial costs, the time required to replenish the inventory, the costs and potential of an inventory stock out, the storage space and costs as well as other factors considered vital to the decision-makers.

Each reason to change a vendor and its effect on documentation will be discussed in detail. This is necessary because improvement techniques will vary for each reason. The documentation that may be effected include the bill of materials, the routing for the product, drawings, the item master file of any new part number required as a result of the vendor change, work order instructions and product quality specifications, installation instructions for the product, and repair and service parts.

An item master file contains at a minimum the assigned part number, the description of the item as detailed as possible, the unit of measure, a notation to indicate if the item is manufactured (M) or purchased (B) and the cost of that item, and additional information that could be included if space is available possible as a specification number; the ECO number that is associated with this item, a revision number of the ECO and the date of its effectivity; and other information as needed if possible. An example of an item master file is below. If the item is purchased, the cost is the purchase price; if the item is a make, then the costs will be calculated based on the bill of materials and the process times and crew sizes from the routing.

Factors to include into the decision-making matrix. This list is not all inclusive.

1. To what degree are other customers satisfied with this vendor?
2. Are the quoted price, terms competitive and comparable?
3. Has the proposed vendor been responsive to requests for samples?

4. How much assistance has the proposed new vendor provided to insure that the new item will perform as needed and satisfy all specifications?
5. What ISO certifications has this potential vendor earned and what is their importance to the decision?
6. What changes in documentation does each motivating factor require?
   a. The replacement of a vendor to obtain an improved price, terms of delivery, or a change in the material used should occur only after assurance that the quality specifications for the new item meet or exceed those of the currently purchased item and the quoted prices, terms of delivery are comparable. A quoted price FOB (free on board) plant is not the same as FOB destination. The cost of delivery at a minimum must be added to a quoted price FOB plant to enable comparable costs comparisons to FOB destination. Other costs that may be included are insurance for the goods in transit and the cost of carrying additional inventory to enable production to continue uninterrupted until the order is received.
   b. If the new vendor is added, then the new purchasing unit of measure must be compared to the existing purchasing unit of measure. If the units are identical, as each, per M, etc., the ECO that initiates the request for change must assign a different part number to this item to enable tracking the date and reason for the change. As the new part number is entered into production, any additional problems encountered in production can be evaluated for increased costs, etc. and can be used to determine if the use of this new item is really cost effective. The material planner will assign effectivity dates for the change in manufacturing, repair, and service parts.
   c. If the purchasing unit differs, then the initiator of the ECO must examine all existing usages of the existing part for the changes in the bills of materials that must be changed to reflect this different purchasing unit of measure. A different part number must be assigned to this item requiring the addition to the item master file that must contain the cost per purchased unit of measure. Manufacturing process times will also be effected and will require studies to determine the extent of the needed adjustments to routings. Additional or fewer manufacturing processes may be needed to accommodate the different purchasing units of measure. All of these changes must be clearly documented in the ECO.

   The different part numbers will enable the calculation of any cost changes that result from the changes in the manufacturing process and waste generation. These costs changes can then be compared to expectations as indicated in the ECO to insure that savings do occur. The material planner will establish an effectivity date based on existing inventory levels, delivery time for the new item, and other variables.

   Another potential reason for a vendor change is a change in the raw material used in the manufacturing process. An example is a furniture company that desires to convert from one type of lumber to another, as from hickory to oak. Conversion from one species of lumber or the grade of lumber used, as converting from #2 to saps or better, may not require a vendor change if the existing vendor is able to supply the desired product at competitive prices.

    d. If a new vendor is required to provide this different material, please refer to the recommended method of selecting a new vendor as listed in item #3 on page Find. A new part number must be assigned to new material, bills of material must be changed to reflect this new material and process routings need to be investigated to determine if process times or the number of operations are affected. These changes must be coordinated and will be effective on the date the planner establishes the effectivity in and out date for the new material. All bills must be changed to include repair and service parts. The assignment of the different part numbers will enable the calculation of cost revisions to insure that the cost savings projected from the change are realized.

       The addition of a new component to be used in the manufacturing process will require the addition of a new vendor if an existing vendor cannot supply this item at a competitive price. If a new vendor is needed, the selection process as detailed above in the solution to question two should be followed.

       The item that is purchased from an existing or new vendor must have a different part number assigned. The ECO will specify at a minimum the documentation affected and the changes that need to be made, the purchasing unit of measure, and the cost per unit of measure.

       The types of documentation affected will depend on the reason for the need of the new item. If used in the manufacturing process, the bill of materials, the process routing, and workplace instructions will need modification. Other documentation affected includes instructions for repair and service parts.

4. What are examples of affected critical documentation?

    a. An example of an item master file (Table 11.1)

       The item master file will contain item number, description, unit of measure, code for make or buy, ECO number and its revision, and other information as desired.

    b. Examples of bills of materials files

       A bill of material consists of the components and the quantities of those components needed to manufacture the product. A manufacturing or process routing is a list of sequential processes that must occur for the product to be manufactured and packaged to meet all specifications. At a minimum, the information included in the routing is the work center that will perform the task, the set up time in hours needed prior to the initiation of the task, the actual time that is required for the standard crew to complete the task, and the standard crew size.

TABLE 11.1

An example of an item master file

Item number	Description	Description	Unit of measure	Make/Buy	ECO number	Revision
001	Back. Sofa, upholstered		ea	M	256	A
002	Plywood, 10 ft × 5 ft × 25 mm	Grade A/D Spec 452	ea	B	256	A
003	Seat. Sofa, upholstered		ea	M	256	A
004	Kit, foam. Sofa		ea	B	256	A
005	Fabric, upholstery	Pattern number stripes	Linear Yd	B	256	A
006	Carton, corrugated	60 in × 24" × 26", 250 lbs test	Ea	B	256	A
XYZ23-455	Sofa, modern style, stripes		ea	M	256	A

1. An example of a simple bill of material is presented below for a sofa that consists of a back produced from a sheet of plywood, a seat also produced from the same sheet of plywood – thus the identical part number for the plywood and a purchased upholstery kit containing the various pieces of foam with the correct size and density. A bill of materials is established with the quantity of units consumed of the incoming product as reflected in the bill to enable pricing and usage to be computed correctly.

    Shippable part number XYZ23455 Sofa, modern style, stripes 1 ea (Table 11.2).

    The foam kit is a purchased item that may contain a number of components. Each of these components must have a different part number assigned since each component is unique. Each component will have a unique drawing number and may have different specifications. Each of these components must be listed in the item master file as a reference part to prevent each component from being ordered. The foam kit is the purchased item, not each individual piece of foam.

    A receipt of an order for this specific sofa would initiate independent demand orders for the first level of components, the back, the seat, and a purchase order for the upholstery kit and carton.

    Two of the independent demand orders would initiate a manufacturing order, one for the back and one for the seat. Each manufacturing order would create a dependent demand for the component needed to manufacture the item and the quantity needed due to the conversion.

2. An example of an indented bill of material to illustrate the demand for items needed to manufacture the major components of a product and the conversion of the number of inputs needed to produce one output is below. The indention indicates dependent demand created as a result of independent demand.

    For example, a sheet of plywood 10 feet (L) by 5 feet (W) by 25 mm, approximately one inch (thickness), is assigned part number 001 at a delivered cost of $56.00 per sheet. The sheet can produce two sofa backs, part 002. The bill of material for the back is below (Table 11.3).

    That same sheet of plywood, 002, is used to produce three seats for that sofa, part 003.

3. The bill of material for the seat is below (Table 11.4).

4. The bill of material for the foam kit, 003, consists of two items which must be coded as reference parts because the purchased item is the foam kit and the demand is for the entire kit and not the individual components. The bill of material for the foam kit is below (Table 11.5).

c. Another document that must be changed is the process routing.

1. The routing of the sofa is shown in the Table 11.6. Other processes that occur, such as inspection, would be shown.

2. A sample routing for either the seat and back can be seen in the table below (Table 11.7).

TABLE 11.2
An Example of a Bill of Material – the Shippable Sofa

Item number	Description	Description	Unit of measure	Make/Buy	ECO number	Revision
001	Back, sofa, upholstered		ea	M	231	A
003	Seat, sofa, upholstered		ea	M	231	A
004	Kit, foam, sofa		ea	B	231	A
005	Fabric, upholstery	Pattern number stripes	Linear Yd	B	231	A
006	Carton, corrugated	60 in × 24" × 26", 250 lbs test	Ea	B	231	A

TABLE 11.3
An Example of a Bill of Material for a Manufactured Component – the Back

Item number	Description	Description	Unit of measure	Quantity	Make/Buy	ECO number	Revision
001	Back, sofa, upholstered		ea	1	M	256	A
002	Plywood, 10 ft × 5 ft × 25 mm	Grade A/D Spec 452	ea	0.5	B	256	A

TABLE 11.4

An Example of a Bill of Material for a Manufactured Component – the Seat

Item number	Description	Description	Unit of measure	Quantity	Make/Buy
003	Seat, sofa, upholstered		ea	1	M
002	Plywood, 10 ft × 5 ft × 25 mm	Grade A/D Spec 452	ea	0.333	B

A work center is an area in which one process occurs. Information for each contains at a minimum an assigned operation number, the normal crew size, its location, a description, and the machine or machines assigned to it. Routing process numbers are assigned in numerical order and refer to the sequence in which the processes occur. It is a good practice when assigning routing numbers to skip a few numbers in case one routing operation or more need to be inserted into the process and to maintain sequential order, such as 050, 100, 150, 200, and so on.

The crew size is required to calculate direct labor cost, which is the product of run time and the direct labor costs of those assigned. The direct labor costs for setup are the product of setup time, the crew size, and the direct labor costs per hour. This product is then divided by the run quantity to determine setup costs per completed item.

Inventory level for the upholstery kit would be checked and the kit may be ordered depending on the current level of inventory, the projected usage over a specified time period, the reorder point which is a function of the length of time the vendor needs to replenish the inventory including ship time.

The independent demand orders for the back and seat would drive a demand for the component needed which is the sheet of plywood. The conversion which is represented in the indented bill of material will result in the number of sheets that will be required to produce the desired quantity of backs and seats. The standard direct labor hours in the routing file and the use of MRP and backward scheduling will indicate the start times for the manufactured components to insure on-time delivery.

The first two items would be coded as a manufactured item and the last item, the upholstery kit, would be coded as a purchased item with a vendor listed with other needed documentation. The relationship between the sofa back and the sheet of plywood required to produce it is often referred to as parent and child, the back being the parent and the plywood, the child.

To illustrate the mechanics of the ECO process, marketing proposes the addition of three identical seat cushions in the same upholstery fabric as the seat and back cushions in order to increase sales. Purchases decides to buy the cushions with the correct fabric ready to install onto the sofa for an interim period. Manufacturing decides to establish a new work center to inspect and install the cushions onto the sofa.

Marketing would initiate an ECO indicating that a new part number is needed for the cushions with the purpose to increase sales. The ECO would state that it was

TABLE 11.5
An Example of the Bill of Material for a Purchased Part – the Foam Kit

Item Number	Description	Description	Unit of measure	Quantity	Make/Buy	ECO number	Revision
004	Kit, foam, sofa		ea	1	B	522	A
007	Foam, back, sofa	Drawing S256	ea	0	Ref	522	A
008	Foam, back, seat	Drawing S285	ea	0	Ref	522	A

TABLE 11.6
An Example of a Process Routing – the Shippable Sofa

Operation description	Work center	Set up time (hrs)	Run time per piece	Drawing number	ECO number	Revision
Cut fabric per pattern	100	0.25	0.325	5466	256	A
Sew seat and back	105	0.08	0.426	17A	256	A
Upholster seat and back	110	0.05	0.751	682	256	A
Assemble sofa	115	0.05	0.335	85n6	256	A
Package sofa	120	0.05	0.15	854	256	A

TABLE 11.7
An Example of a Process Routing for a Manufactured Component

Operation description	Work center	Set up time (hrs)	Run time per piece	Drawing number	ECO number	Revision Level
Rip plywood to width	200	0.05	0.022	583a	256	A
Cross cut to depth	205	0.06	0.015	547	256	A
Sand all four sides	220	0.08	0.126	954	256	A

a purchased item to be installed after the sofa assembly operation. The ECO would include several price quotes from different vendors, the new work center number with the crew size, and a time expressed in hours for installation and distributed through the various affected functions. If the ECO was approved, then purchases would employ the vendor selection procedure to determine the vendor. Engineering would add the new part number to the item master file, add the cushion, quantity of three since there are three cushions per sofa, to the bill of material to the sofa, add the new work center to the work center file, and add the operation to the routing for the sofa. Material planning would add the effectivity in date during the approval process (Table 11.8).

The revised bill of material for the sofa appears in Table 11.9.

The revised routing for the sofa is below (Table 11.10):

Often the industrial engineer will have the opportunity to work with a salesman to develop new opportunities for existing products, to modify processes to better fit customer needs, or to recommend different products that satisfy customer needs with one that is lower priced. The author has experienced this opportunity on numerous occasions for different products, different customers, and different product uses. When given this opportunity for his company to become a vendor, this provides the opportunity to learn the attributes of a good vendor which can then be applied to his company to determine improvements in the service it provides to existing customers and to increase its customer base.

TABLE 11.8
Bill of Material for the Sofa before the Addition of the Three Cushions

Item number	Description	Description	Unit of measure	Make/Buy	ECO number	Revision Level
001	Back, sofa, upholstered		ea	M	256	A
002	Plywood, 10 ft × 5 ft × 25 mm	Grade A/D Spec 452	ea	B	256	A
003	Seat, sofa, upholstered		ea	M	256	A
004	Kit, foam, sofa		ea	B	256	A
005	Fabric, upholstery	Pattern number stripes	Linear Yd	B	256	A
007	Cushion, sofa, stripes	19 in × 26 in	ea	B	257	A
006	Carton, corrugated	60 in X 24" × 26", 250 lbs test	Ea	B	256	A
XYZ23455	Sofa, modern style, stripes		ea	M	256	A

TABLE 11.9

Bill of Material for the Sofa with the Addition of the Three Cushions

Item number	Description	Description	Unit of measure	Quantity	Make/Buy	ECO number	Revision level
001	Back, sofa, upholstered		ea	1	M	256	A
002	Plywood, 10 ft × 5 ft × 25 mm	Grade A/D Spec 452	ea	1	B	256	A
003	Seat, sofa, upholstered		ea	1	M	256	A
004	Kit, foam, sofa		ea	1	B	256	A
005	Fabric, upholstery	Pattern number stripes	Linear Yd	5.67	B	256	A
007	Cushion, sofa, stripes	19 in × 26 in	ea	3	B	257	B
006	Carton, corrugated	60 in × 24" × 26", 250 lbs test	Ea	1	B	256	A
XYZ23455	Sofa, modern style, stripes		ea	1	M	256	A

TABLE 11.10
The Process Routing of the Sofa after the Addition of the Three Cushions

Operation description	Work center	Crew size	Set up time (hrs)	Run time per piece	Drawing number	ECO number	Revision
Cut fabric per pattern	100	1	0.25	0.325	5466	256	A
Sew seat and back	105	1	0.08	0.426	17A	256	A
Upholster seat and back	110	1	0.05	0.751	682	256	A
Assemble sofa	115	2	0.05	0.335	856	256	A
Inspect and install cushions	125	1	0.03	0.07	8922	257	B
Package sofa	120	3	0.05	0.15	854	256	A

C. CRITICAL QUESTIONS THAT MUST BE ASKED BEFORE INTEGRATING MEASUREMENTS INTO THE CHANGING OF A EXISTING PROCESS

1. Is there a change in the unit of purchased unit of measure (u/m)?

   A change in the u/m will affect the bill of material, the process routing, the manufacturing costs, drawings and possibly the receiving process.

2. Is there a packaging change for the incoming product?

   A packaging change will affect the incoming receiving process, the material handling of the material and the raw material inventory procedures.

3. Is there a change in the packaging specifications?

   This will affect the QA inspection procedures, drawings, process instructions, and process routings.

4. Is there a change in the FOB point?

   This will affect the purchased cost of the product. If the change is from FOB destination to FOB plant, then the transportation costs must be added to the purchase price.

5. Who is responsible for initiating the ECO and assuring the accuracy of the change in affected documentation?

   The initiator of the ECO is the person requesting the change. For example, if engineering initiated the change they are responsible for initiating the ECO and ensuring that all affected functions are included and for the accuracy of the changes. If QA initiates the change, they are responsible with the assistance of engineering to ensure that all affected functions are included and that engineering is responsible for the accuracy of the changes.

D. EXAMPLES OF SUCCESSFUL AND UNSUCCESSFUL VENDOR CHANGES

1. Successful vendor changes
   a. An example of a successful vendor change in which the author's company became a vendor for a new customer resulted from a phone call

while the author was in a company plant. The potential customer was a furniture plant that was seeking an alternative to the current product it was using to protect a hanger that was used in the finishing department. A salesman and the author called on the company to determine what were the needs and how best they could be met.

The current product being used was a spiral tube 1½ inches in diameter (think of a tube on which paper towels are wound) that was 48 inches in length with a saw cut that was parallel with the length of the tube to enable the tube to be opened as a hot dog bun. The purpose of the tube was to surround a hangar that connected the bracket on which furniture was placed for filling and staining as the overhead conveyor as the furniture was transported through the finishing department. The thickness of the spiral tubes was 0.125 inches. The current vendor was located several states distant. The tube protected the bracket from accumulating excess finishing material which could become a potential fire hazard due to flammability of the finishing materials. At least annually the paper tube was replaced.

On the return trip to the plant it was decided that a convolute tube, think of a straw, would satisfy the needs of the customer. A convolute tube is less expensive to manufacture due to fewer processes involved. Also, after returning to the plant a check with maintenance confirmed that a device similar to a pizza cutter could be installed after the tube is ejected from the mandrel resulting in no additional costs incurred for this process. A minimal upcharge was added to the price of the tubes to recoup the maintenance costs for making and adding the cutting device onto the machine.

A sample of 50 was produced, a price was computed, and both were personally delivered to the customer within a few days. The foreman of the finishing department, salesman, and author took the samples to the finishing department and met the head of the safety department. Both were satisfied with the product and price and the salesman left with a purchase order for a T/L due the following Friday to enable maintenance to change them over the weekend. The tubes were packaged as requested, 19 per bundle, and hand loaded into a trailer to prevent tubes from crushing while being packaged and shipped.

This opportunity for the author's company to become a vendor was a success because the following critical questions were asked and answered:

1. What was the current product being purchased?

     A spiral tube 1½ inches in diameter, 48 inches in length, and 0.125 inches long with a slit cut down one length enabling the tube to be opened like a hot dog bun.
2. How was it being used?

     The tube after it was opened was placed around a hanger in the finishing department to collect overspray to prevent the overspray from accumulating in the paint booths reducing cleanup time and potential fire hazards.

3. What were its specifications and how critical were they?

The only critical specifications were the length of tube because that was the length of the hangar and that the slit in the tube must be complete.

b. Another successful example of an industrial engineer who worked with a salesperson that yielded additional a new customer that was extremely profitable occurred when the author made a sales call with a salesman to see a packaging engineer who had inquired if the company could supply what he termed a core plug (Figures 11.1 and 11.2).

The purpose of the plug was to secure the stretch film used to package the final product by inserting the core plug into both the ends of the tube onto which the material was rolled after the stretch film was applied. The core plug had a ½ inch lip which was a paper tube stapled onto one end of the core plug to prevent slippage inside the paper tube before the placement of a shrink bag over the entire pallet which was then shrunk to ensure package integrity.

The core plug consisted of a paper tube with a diameter small enough to fit inside the tube onto which the product made was wrapped by the company and had a larger tube stapled at one end that acted as a stop. The thickness of both the tubes was measured from the existing vendor and found to be 0.500 inches. The tube acting as stop was stapled onto the other tube.

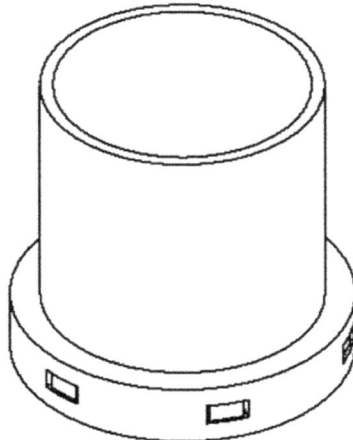

FIGURE 11.1   Core plug used to insert into end of rolled goods after installation of outer packaging to maintain package integrity (orthogonal view).

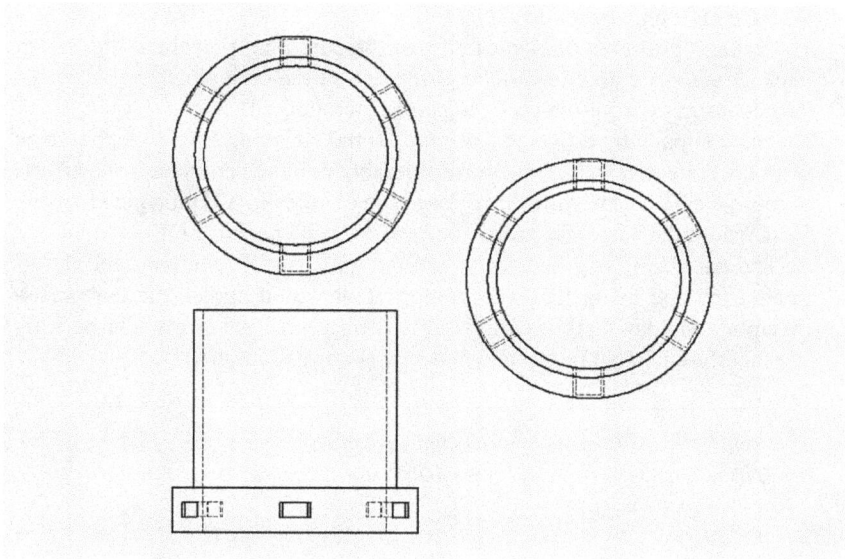

FIGURE 11.2   Core plug used to insert into end of rolled goods after installation of outer packaging to maintain package integrity (top, plan, and right side view).

The goal of the packing engineer and the PA was to consolidate vendors. Another advantage was a consolidation of vendors since the author's company would now supply the larger tube used to wrap the material and the two core plugs.

The salesman and engineer followed procedure:

1. Samples of the current product were obtained and tested.
   The only tube that was evaluated was the one inserted since it was subjected to pressure.
2. Samples were provided – a staple machine was located and borrowed for a trial.
3. A quote was provided for the core plug.
4. The samples were evaluated in production by the QA, the production engineer, and the author.
5. All agreed that the performance of the samples was acceptable.
6. The salesman provided the PA with the quote who responded with a purchase order for a T/L order due the following Monday.

The critical questions that were asked and answered that resulted in this acquisition of a new customer are as follows:

1. Was the core plug necessary?

    Packaging test results indicated that the core plug prevented the loss of package integrity and thus the requirement for the core plugs.

2. What alternatives for the core plug were investigated?

    The company investigated several options, including PVC pipe, but none were cost effective and all were considerably more expensive despite efforts to recycle a PVC pipe with a lip of one inch of glue of PVC pipe glued on end

3. How would the customer prefer the core plugs be packaged?

    The packaging engineer met with the manager of receiving and developed a method using RFID technology that would update the customer's inventory and paying the vendor within the agreed upon terms. The packaging method included recyclable material (Figures 11.3 and 11.4).

Title	Palletized core plugs (orthogonal view)				
Figure number	11.3	Drawn By	JD	Date	9/26/2020
Palletized material					
Checked by	JW	Date	2/21/2021	ECO #	928-021

2. Examples of unsuccessful vendor changes

    a. This is an unsuccessful effort in an attempt to obtain a new customer that produces paper with multiple uses such as for copying or printing documents. The effort did result in the author's acquiring a new vendor but at a loss. This example is labeled a failure due to the loss incurred as a result of the additional sales volume.

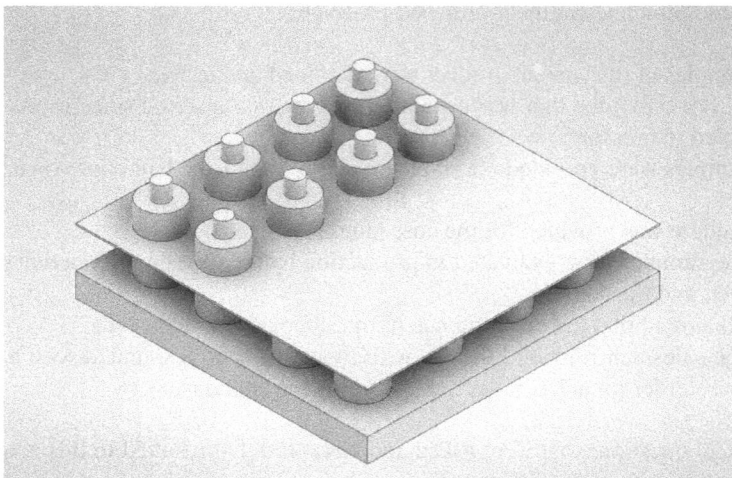

FIGURE 11.3   The packaging procedure for core plugs (orthogonal view).

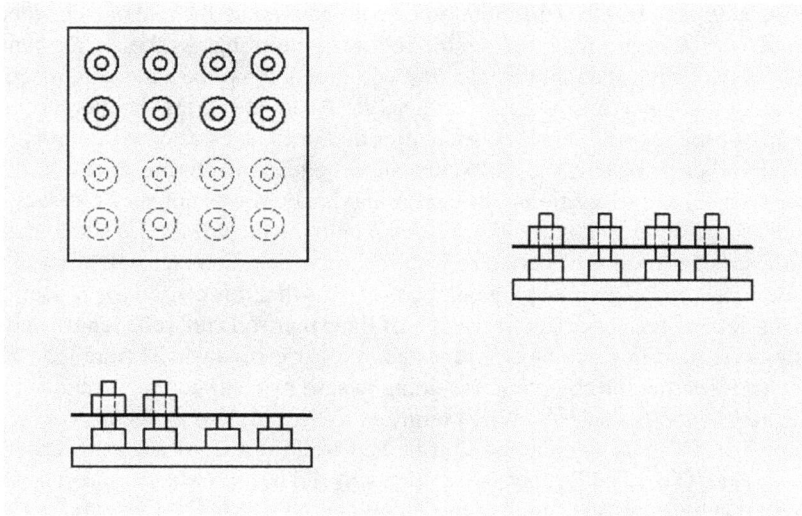

FIGURE 11.4   The packaging procedure for core plugs (top, plan, and right side view).

b. The author and a salesman made a sales call to a plant that manu-
factures paper used in copy machines. This is a large company with
manufacturing plants nationwide. The purchasing agent provided the
salesman with detailed specifications consisting of seven sheets which
the salesman quickly looked through and responded that there will not
be a problem in providing a product that met or exceeded these specifications. The salesman quickly calculated the selling price which he
quoted to the PA and looked the PA in the eye as they shook hands. The
sales call was made on a Tuesday; the purchasing agent stated that they
would need a truck load of tubes on the following Monday. The salesman asked if there was a specific time that the delivery should be made.

After examining the specifications, the author realized these tubes would need to be
produced in another plant due to the additional processes required. The additional
freight or the cost of the additional processes had not been added into the price
quoted to the PA. The author did not question the salesman since to do so at that
time would only have undermined the salesman. Those questions should have been
addressed and resolved if the specifications had been made available to the salesman
and the engineer prior to the appearance in the PA's office. However, given the situation if the purchasing agent had asked the author, the author would have suggested
that the company supply a 1,000 M test samples for a test run to be monitored by
the author, the company's engineer, and QA, to determine how well our product performed and make recommendations as to the next step.

The truck load of tubes was shipped and received the following Monday. The
author received a call from the purchasing agent within several hours of receipt (the

receiving company was in a different time zone) who asked the author to be there in his office with that salesman at 8 am the following morning. As the author hung up the phone, the author thought this meeting will probably not be a pleasant meeting.

Arriving at the purchasing agent's office, the PA indicated that he had received a call on that Monday from his boss soon after the tubes had started production just a few minutes before placing the call to the author requesting our presence the following morning. The conversation with the PA and his boss was not very pleasant. The PA then escorted the author and the salesman onto the plant floor. He stated that the tubes the company shipped varied in length and outside diameter (OD) by as much as 0.010 inches and explained in detail the problems that this caused to production in the plant. Due to the speed and the width of the paper roll run, tube length and OD variations resulted in machine stoppages and cost overruns which more than offset the few pennies that the company was going to save by changing tube vendors.

The author and the salesman then returned to the purchasing agent's office. The PA looked at the salesman and asked if he had read the specifications to which the salesman said "No." The PA then asked then why did he ship what he did. The salesman said that his company supplied the PA's plant in Dallas. TX; this plant in Dallas was also making copy paper and that the specifications for that plant were only two pages. The PA quickly stated that there was a reason that their plant had much more detailed specifications and that if the salesman had taken the time to read them he would have found out what the difference was. The PA also stated that this plant was in Alabama and not in Dallas. At that the author thought the PA was going to scream but instead he calmly asked the author what would the author have done after having examined the specifications. The author responded that he would offer a sample of 1,000 M tubes to be monitored for results by the author, the company's engineer, and QA, to determine how well the product performed and make recommendations as to the next step.

The PA then calmly called the president of the author's company and told him in no uncertain terms that he never wanted to see that salesman again. He asked the president three questions:

1. Can the company meet the specifications? To this the president deferred to the author. The author responded possibly but tubes would need to be verified by testing and the only possibility of meeting those tight specifications would be if the tubes were produced in a plant two states distant.
2. Will the company honor the quoted price? To this, the president agreed.
   The president agreed to meet the quoted prices although he suspected that the items would be produced at a loss. The president emphasized to the PA that honesty, integrity, and service were more important in the long run than a few dollars of potential profit.
3. Does the company want the business? To this the president indicated that he did.
   The PA then told the president that the only person he wanted to see was the author. The president then told the PA to work out the details with me. He then asked how soon could samples be shipped. The author

responded that at this time it was not known since the company must first ensure that it can guarantee that it can manufacture the product to meet or exceed the customer's specifications. The PA would be advised of the progress.

This attempt to acquire a new customer was eventually successful because the following critical questions were asked and correctly answered:

1. Why were these specifications so detailed and what was the difference between these and the specifications for the plant in Dallas?

   Although both plants manufactured copy paper, the end uses of the paper were entirely different requiring more stringent manufacturing processes and quality control during manufacturing.

2. Why were not samples taken for testing and analysis?

   The salesman was in a hurry to get the order and bypassed standard procedure which is to:

   a. Obtain samples from the current product being used for testing.
   b. Observe the current product in actual use.
   c. Compare the test results with the specifications.
   d. Provide a quote FOB plant.
   e. Provide a sample of a minimum of fifty tubes to the customer that meet or exceed the specifications.
   f. Observe the provided samples in operation with all effected functions present to validate that the samples met the minimum criteria.
   g. Follow up with the PA, the engineer and QA.

3. Was the proper procedure for acquiring a new customer followed?

   A phrase attributed to Benjamin Franklin, "Haste makes waste," is actually paraphrased from Proverbs 21 [52]. A recent publication that summarized mistakes and developed a strategy to reduce them concluded that 15.2% resulted from the decision-maker being in a hurry to make the decision, as if the decision-maker had something more important to do than that which he was getting paid to do [53].

   If the salesperson followed standard procedure in acquiring new customers was followed resulting in the following actions.

   Samples of the current product being used were obtained and tested.

   a. Observe the current product in actual use to determine the existence of any problems and their resolution if any were noticed.
   b. Compare the test results with the specifications which are needed for the preparation of a quote.
   c. Provide a quote FOB plant of a product that meets or exceeds current specifications or add freight if the mileage to the customer exceeds the shipping range.

The tube in order to meet the tighter customer specifications would have to be manufactured in a plant several states distant from the customer that would require the

incurring of additional freight upcharge. Additionally, due to the additional processing required, the tubes would need to be packaged differently. The new package
design would require a unitized package with recyclable material that would retain
moisture to prevent length, diameter, and strength variations changes. The unitized
package incorporating the use of shrink film was suggested by the customer due to
the extended delivery time.
Within a month that company was an excellent customer. The successful acquisition
of this new customer resulted in a net loss for the company for the following reasons:

1. The product required manufacturing at a different plant, incurring additional freight charges that were not included in the original quote.
2. The product required additional processing which due to equipment availability had to be done at another facility. This additional processing or the
   freight from the distant facility was not included in the original quote due
   to the salesman's desire to deliver a quote immediately.
3. The president of the company guaranteed the quoted price for a year during
   a phone call with the purchasing agent. The president did indicate that the
   price was subject to change after the year but as a result of the failure of the
   salesman to follow protocol, the price that was quoted would be held for one
   year regardless of volume sold.

In a previous text "Critical Thinking: Learning from Mistakes and How to Prevent
Them" [53], the author concluded based on his own personal experience that over
80% of the mistakes he committed and witnessed could have been prevented if the
decision-maker had taken a few minutes to take an additional step before making a
decision. Examples of extra steps include this extra step that could have eliminated
over 80% of the mistakes that occurred had he taken some appropriate timely action.
For an industrial engineer an opportunity such as this is similar to winning the lot-
tery even if he had to pay taxes. This is the low hanging fruit that all industrial
engineers seek (Figure 11.5).

In retrospect, had the author simply followed his own advice and gotten an additional piece of information, many dollars, jobs, and much time would have been
saved. The acquisition of one additional piece of information to prevent or reduce
the severity of errors by the employment of critical thinking is the approach the Joint
Commission employed to address wrong site surgical procedures. The Commission
issued a protocol for preventing wrong site, wrong procedure, wrong person surgery.
This protocol was comprised of several steps:

1. Verification of the correct patient, documents, site, etc.
2. Marking of the operative site prior to the operation with the agreement
   of the patient so that mark will be visible after the patient is prepared for
   surgery.
3. Taking of a timeout to increase teamwork before surgery which has been
   shown to reduce the number of incidences of wrong site surgery. The marking of the site that provided an additional information for the surgical team

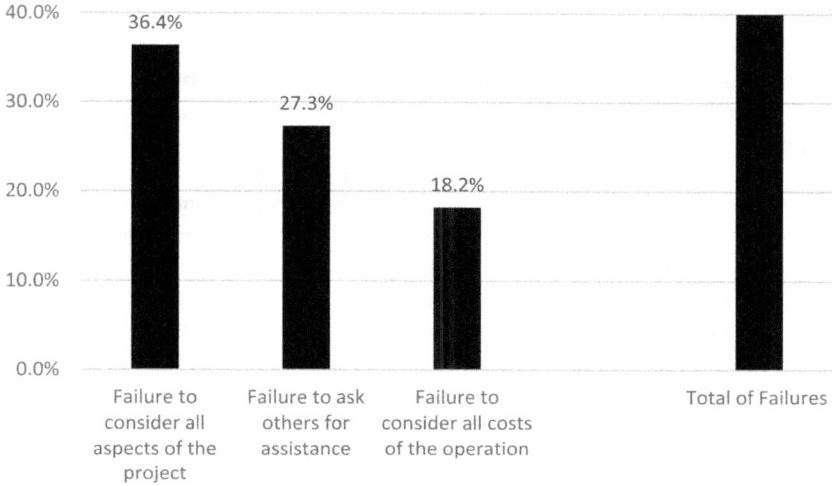

FIGURE 11.5   Graph illustrating that over 80% of mistakes result from the failure of the operator to take one additional step before taking action.

to eliminate and reduce the likelihood of an error and the increased team-work provided by the timeout are two of the many human factors which assist in the mitigation of mistakes [55].

The question that must be asked is what if there is a difference between the specifications provided by the vendor and those determined from the samples received from the customer.

As one who has encountered this exact situation numerous times, the author has returned to the customer and asked for a few additional samples for testing. One always speaks highly of competitors. Regardless of the outcome, the author has always provided the customer product that met or exceeded the provided specifications. Only after he became a customer has testing begun with customer approval to reduce specifications, if possible, for reducing cost.

Another example of unsuccessful a vendor change – tubes to package airplane tires

   c. The tires were packaged in a carton and shipped via railcar and used spiral wound tubes. Think of the tube on which paper towels are wound, because it has some crush strength side-to-side and some end-to-end strength in the corners of cartons to provide support as these cartons were transported in railroad cars.

     While talking with the PA the author discovered that the next order was due in one week. The self-imposed arbitrary deadline to receive the next full order to secure this account for the company drove the author to bypass procedures. Normal procedures were to test the existing product, provide a sample of the proposed product as well as one that equaled the one being currently used with a delivered price for each, and then provide an opportunity for engineering and QA to perform testing to determine what product performed best.

Upon returning to the office, the author with the help of the salesman determined the mileage, calculated the freight upcharge and added that to calculate a delivered price and provided it over the phone. The salesman was pleasantly surprised when the PA told the author and salesman that the quoted price was very attractive.

This attempt at securing a new customer failed because the established procedures were not followed. The tubes that were supplied did not meet or exceed minimum specifications resulting in the cartons in the railroad cars being tossed around during shipment and the spiral tubes which were able to endure the side-to-side to side pressure helped the cartons to remain intact. Conversely, the convolute tubes were not able to withstand the side-to-side pressure resulted in the cartons collapsing and releasing the airplane tires within the railroad cars. This required that the railroad cars be unloaded manually as opposed to the use of automatic equipment as fork trucks. The salesman and the engineer received several phone calls from people who were not happy after having to unload numerous railroad cars manually. These unhappy phone calls resulted in unpleasant long discussions with our boss.

d. An example of an attempted vendor change that initially failed but was finally successful after the loss of money and prestige.

This loss resulted from a purchasing agent failure to ask the critical questions concerning a difference in price when given a quote of a product used by this company:

1. Why was the price less expensive?
2. How did this product compare in QA tests with the existing one?
3. What effect would the supplier being offshore have on the supply chain?
4. Who pays the freight to the plant that will be using the caster?
5. Has the current vendor been given an opportunity to meet the new quote?

The item in question is on the bottom of an office swivel chair, a caster. There are five per chair. At one time there were four but these chairs were easily tipped over and could no longer be produced. Materials and/or processes by which a chair is processed must be approved by all involved including marketing, purchasing, engineering, quality, and all levels of management The process by which these changes occur is the engineering change order system; in most companies, with multiple plants this system is electronic. The author received an ECO to change from the current vendor to one in a different country. The cost difference was $0.01 per caster. The caster was priced FOB manufacturing plant which meant that the company for which the author worked was responsible for paying the freight from this foreign country to the facility for which the author worked. This compared unfavorably with the terms of the current vendor which was an FOB facility.

All the other plants in this company signed the ECO signifying agreement with the change. The author did not sign the ECO and indicated the following reasons.

1. There was no attached documentation indicating that the caster from the new vendor was at least as good quality-wise as the existing caster.
2. There was no documentation indicating discussion had occurred with the existing vendor that a competitor had provided a lower quote and the existing vendor had been given an opportunity to meet or exceed it.
3. The company that employed the author had been conducting business with the current vendor continuously for many years with no quality or delivery concerns.
4. In the opinion of the author as much manufacturing as possible needed to be retained in the United States as possible.

The purchasing agent was not happy with the response but appreciated the honesty. Several weeks later the author received a call from a fellow engineer from a sister plant in another state asking if the plant at which the author worked had any casters from the original vendor. The author responded that the plant did but when asked why, the engineer stated that the first shipment from the alternate vendor passed inspection and performed as expected but when the cartons from the second shipment were opened, all that could be seen were plastic components. The casters apparently broke apart in shipment. The author checked with the material planner in his plant to determine how much, if any could be shipped to help a sister plant. The material planner called the fellow engineer and told him that the plant would ship as many as possible without jeopardizing that plant's shipments. The plant at which the author worked placed an order with the original vendor for as many casters as possible to arrive as soon as possible.

The cost savings or a penny per caster does not seem to be much but one must first be aware that there are five casters per chair. This company produced over 6,000 chairs per day in their different chair plants so a change was justifiable if the proper conditions were met. However, this mistake resulted in the inability to ship chairs for many weeks resulting in many lost dollars and prestige which cannot be measured in dollars. The author would not want to explain to a customer that his chair could not be shipped due to the desire to save a penny a caster. The only plant that did continue to produce and ship chairs without interruption was the one that did not change caster, which was the one at which the author worked. The motivations of this plant to not approve the change was that the freight costs from the new vendor and the cost to carry the needed inventory to cover the additional travel time would exceed the penny difference in the cost of the casters.

e. An example of a failed vendor change involved a furniture manufacturer of a traditional line of bedroom and dining room furniture. The plant manager intended to change vendor to one that supplies a different lumber to produce the unseen wooden components in the furniture from whatever was currently being used to sweet gum. Sweet gum was less than half the price per thousand board feet of the least expensive grade of lumber than was being used at the time.

As an engineer the typical questions were asked if there would be additional maintenance costs and machining costs and was assured that the price difference would more than compensate for whatever incremental manufacturing costs changing to sweet gum may incur. He mentioned the need to change vendors due to the few vendors offering sweet gum in the sizes used. This company had been in operation in a small town in the heart of the furniture industry for many years.

However, soon after the company changed vendors and began purchasing lumber from different vendors for those components that the customer does not see. The manufacturing costs began to rise significantly due to waste and the amount of product that had to be reworked due to the splitting of the sweet gum upon the introduction of a staple and a large number of unseen splits. In addition to difficulty in splitting, other problems exist with sweet gum. Although there is an ample supply, little is seen in sawmills or in furniture plants. One problem is that it shrinks, or warps in all directions because the material has interlocked grain [56].

In addition, due to the greater density in comparison to its replacement, the tools used in the manufacturing process required more frequent sharpening and replacement, adding additional costs to manufacturing. Within a year this decision resulted in the closure of this plant and the loss of hundreds of jobs.

This failure was due to the lack of critical thinking. If the following questions had been asked and answered, the change would never have occurred and the company would most likely still be in operation.

1. If sweet gum is such a good substitute, especially at such a low price, why is it not more widely used?
2. Will the sweet gum be tried on a trial basis to determine feasibility?
3. Why weren't sample products produced to determine manufacturability before the conversion was made?

f. Another unsuccessful example of a vendor change involved a motorcycle manufacturer that changed plating vendor.

Measurements that were not taken by the manufacturer but contributed to the failure of his once-thriving enterprising business.

1. The total of the components that were plated from current vendor including freight, inventory, and waste.

2. The total cost of plated costs from proposed vendor including freight, inventory, and waste.

    The realization that a disruption in supply or an increase in waste would jeopardize his business to the point of bankruptcy and its elimination never entered the owner's mind due to the reliability of all the current vendors including the plater.

    Measurements that were not taken by the manufacturer but contributed to the failure of his once thriving enterprising business.
3. The total costs of plating from current vendor including freight, inventory, and waste.
4. The total cost of plated costs from proposed vendor including freight, inventory, and waste.

    The realization that a disruption in supply or an increase in waste would jeopardize his business to the point of bankruptcy and its elimination never entered the owner's mind due to the reliability of all the current vendors including the plater.

12 Integrating Measurement in the Change of a Vendor

A. BRIEF OVERVIEW OF THE NEED FOR MEASUREMENT IN THE CHANGE OF A VENDOR

There are multiple reasons to change a vendor including a deterioration of service or in the quality of products supplied, a vendor shutting, the introduction of a new product that will require materials that existing vendors are unable to provide or can offer at competitive prices.

Similarly, there are reasons based on the experience of the author that will be augmented with examples not to immediately change vendors, especially ones who have provided excellent service at competitive prices over a long period of time. This reason is if a potential vendor offers a currently purchased item at a lower price, there must be a reason that this new vendor can offer the same quality product for a lower price. The company, in the author's opinion, has an obligation to the current vendor to first verify that the quality levels and delivery terms are identical and then offer the current vendor an opportunity to match the offered price. The author has witnessed many situations in which a company attempted to save money by changing vendors that offered the lower price only to suffer huge losses due to the new vendor after providing samples that were approved. After approval the vendors began to ship defective products or ship late causing the company to be unable to meet promised ship dates resulting in the loss of prestige and eventually customers.

B. CRITICAL QUESTIONS THAT NEED TO BE ASKED IN THE SELECTION OF A NEW VENDOR FOR AN ADDITIONAL ITEM, PRODUCT OR SERVICE WITH SUGGESTED SOLUTIONS

1. Have current vendors been asked to quote this new item?

This should be the initial step to acquire this new item. The relationship between vendor and customer must remain transparent. Should a current vendor be unable to supply the new item at a competitive price and terms, then the customer should seek another vendor.

DOI: 10.1201/9781003153412-12

C. CRITICAL QUESTIONS THAT NEED TO BE ASKED IN THE DECISION TO CHANGE VENDORS OF A CURRENTLY PURCHASED ITEM

1. Is the vendor bankrupt or has discontinued operations?
2. Has the vendor changed product lines?
3. Has the vendor become unreliable?
4. Have you used the vendor selection matrix as discussed in Chapter 11?

D. EXAMPLES OF SUCCESSFUL AND UNSUCCESSFUL INTEGRATION OF MEASUREENNT INTO THE SELECTION OF A NEW OR REPLACEMENT VENDOR

1. Examples of successful integration
    a. In the successful example discussed in Chapter 11 concerning the slit in the convolute tube, measurements that needed to be taken were the number of defective tubes that the slitting operation generated and the effect adding the slitting operation had on direct labor costs. Studies were taken after installation of the slitting operation that resulted in no effect in the percentage of defects or the direct labor costs.

        These measurements provided information from which management and engineering concluded that no costs increased from the manufacturing process of slitting of the tubes. Since the costs of adding the slitter knives and the cost of maintenance of the knives are minimal, management agreed upon a minimal upcharge. An upcharge was justifiable due to the change in the specifications required.

    b. Another successful example of integration resulted in a new customer

        Measurements taken included determining the incremental direct labor costs of stapling the two tubes together and packaging the tubes to meet specifications.

        These measurements were taken via formal time study.

        The process of manufacturing this core plug consisted of stapling a recut tube on the end of a larger tube to act as a stop. The information that was needed to compute the sales price was the cost of packaging and the direct labor costs of stapling the small recut tube onto the end of the larger tube. The material costs of packaging were retrieved from recent invoices from the vendor that supplied the strapping.

        A formal time study was taken of this operation that consisted of a crew of one. The steps in the process were to remove a larger tube from a carton containing the large tube, placing it onto the stapler, removing a recut tube from a carton adjacent to the staple, and aligning the recut tube onto the larger tube, activating the pneumatic stapler by depressing a foot pedal, rotating the assembly on the arm of the stapler approximately three inches, and continuing the process around the larger tube and then placing the completed assembly onto the corrugated sheet used for packaging. The core plug will have seven staples approximately equally spaced around the circumference of the recut tube. The time study included replacing the two cartons and operating the stretch wrap machine that assured package integrity.

Direct labor costs have been defined as "the wages or salaries paid to employees who physically produce products" [57]. Included in direct labor costs are the wages paid to employees, taxes, the cost of workman's compensation, and life and health insurance [58].

The direct labor hours to produce a product are the products of the direct labor hours per unit and the crew size. The labor variance is the difference between the actual cost per unit and the standard cost unit.

The difference between the actual costs per unit and the standard costs per unit is the labor variance. A negative variance means that the actual costs are greater than the standard and needs to be studied to determine the reasons that the production standards are not being met. If the variance is positive, it means that standards are being exceeded and studies need to be conducted to determine the causes.

This time study provided the standard direct labor hours for the assembly process.

The standard direct labor costs were then computed per M tubes. The cost of the packaging material was calculated and the costs per package were converted to packaging costs per M tubes. Overhead including a profit percentage was allocated to these costs. The existing selling prices for the larger and the recut tubes were determined and added to the total assembly and packaging costs to obtain the sales price per M tubes.

2. Unsuccesful examples
   a. An attempt to supply a company that produced copy paper resulted in failure.

   The measurements needed to increase the probability of obtaining a new customer were the OD and length in the tubes produced for a different plant of that company, other measurements as beam and crush strength, and moisture content. Several samples of tubes currently in use needed to be obtained and the above measurements from the sample tubes.

   This information could then be used to determine if the tubes supplied to the different plants would perform satisfactorily.

   The company did acquire the customer but at a loss due to the failure of the salesman to follow standard procedures before providing the customer a quote for a new product. The specifications required that these tubes be produced at a different plant due to a tighter length tolerance requiring a freight upcharge, that the tubes be in a shrink-wrapped package to maintain the moisture level to minimize changes in length,

inside and outside diameter after manufacture due to moisture variations. This shrink wrapping operation would represent an additional
expense.

The engineer determined first the standard costs of this item.

The quoted price per M tube as per the salesman was assured to
be valid for at least one year by the president since the salesman had
looked the PA in the eye and shaken his hand when given the quote
verbally.

The cost of the new materials exceeded the carton costs plus the
direct labor for inspection and packing the carton.

The result was a loss in contribution to overhead.

The reason that the company knowingly suffered this loss was that
the salesman had given his word and shaken the hand of the PA at the
time the quote was given verbally. Also when the president was asked
by the PA if the quote would be honored for the terms of the quote
which was a year, the president without hesitation answered yes since
the reputation of this company companies and those in it meant more
than the few dollars than the company would lose. Also, the company
had valuable information that could be used should it be needed if a
company had a situation that had a requirement for similar tube demand
of providing assistance that failed due to not following established
procedures

   b. An failed example of integration was at a company that produced airplane tires

This application failed as a result of the failure to follow proper
procedures.

Measurements that should have been taken:

1. Crush end-to-end tests (beam) tests of the product being used – a spiral
   tube.
2. Crush end-to-end tests (beam) tests of the proposed product – a convolute
   tube.
3. A delivered price of the current price being used.
4. A delivered price of the current price of the product being proposed.
5. An analysis of the crush tests to determine the applicability of convolute
   tube.

The failure to take these tests resulted in the loss of potential customer.

   c. Another unsuccessful example was the attempted change of a caster
Measurements must be comparable and consistent. There are several
legitimate issues with this attempted conversion to another vendor:

1. There is no documentation to demonstrate that the caster offered by the new
   vendor is equal to or exceeds the specifications of the existing one.

2. The price for the new vendor does not include freight, the caster cannot be used unless it is in the plant in which it is to be installed on the bottom of chairs.

    d. Another example of an unsuccessful vendor change occurred due to the failure to take needed measurements.

        Motorcycle manufacturer that changed plating vendor.

    Measurements that were not taken by the manufacturer but contributed to the failure of his once-thriving enterprising business.

1. The total costs of plated costs from the current vendor including freight, inventory, and waste.
2. The total cost of plated costs from proposed vendor including freight, inventory, and waste.
3. Information from customers of the proposed vendor concerning quality and delivery issues, the frequency and severity of these issues, and the length of time that this vendor has been providing quality products.
4. The realization that a disruption in supply or an increase in waste would jeopardize his business to the point of bankruptcy and its elimination never entered the owner's mind due to the reliability of all the current vendors including the plater.

13 Integrating Methods Improvement into Existing or Establishment of a New Safety Program

A. BRIEF OVERVIEW OF THE NEED FOR AN EFFECTIVE SAFETY PROGRAM

The benefits of an effective safety program to an organization is the development and sustainment of a safety culture, a reduction in the accident and incident rates which will cause a reduction in the costs of workers' compensation insurance, a reduction in direct labor costs from an increase in production, a reduction in the absenteeism rate, improved morale, and a lower turnover rate.

B. CRITICAL QUESTIONS TO ASK TO ESTABLISH A NEW SAFETY PROGRAM WITH SOLUTIONS

1. Are top and all levels of management fully committed to the need for the safety program?

   Without management support, the safety program or any other program will have limited success. Management must provide resources as needed to insure the success of this or any program.

2. Does the management provide the ability of a safety team member to stop an operator from continuing to perform a task if the safety team member opines that the manner in which the task is being performed is in an unsafe manner?

   This level of support is essential to enable the safety team to achieve its goals. The solution is to have at least one more than required fully trained and capable operator who can safely operate each machine or perform each process. Management must enforce this rule by restricting the performance of operations and processes to those operators who are fully trained even if production is delayed. The avoidance of delaying production cannot be used as an excuse to allow a less than fully trained operator to perform the task. The solution is to train more operators.

3. How broad a scope will the safety team have?

DOI: 10.1201/9781003153412-13

The scope will determine the members and activities of the safety team. For example, will office personnel be included, all manufacturing personnel to include those involved in indirect labor, as receiving, and inspectors and shipping, If the scope involves training, the team must decide the specific training, who will receive the training, the frequency and method, and who will provide the training. The safety team may include a member who can provide some training as forklift training or in simple stretching exercises to minimize strains.

4. How will the members of the safety team be recruited?

Employees who volunteer to serve will become more engaged and will make greater contributions than those employees selected by other methods. Safety team members should represent different areas of the organization and rotate periodically.

5. What will the initial goals of the safety team be?

a. To educate team members of the current rates of the accident and incident rates of all employees.

This information is available from the OSHA logs which must be posted in a prominent place for maximum visibility.

b. To identify hazards which is any source of potential damage, adverse health conditions, or harm on anyone or thing, as a machine. The team will perform a risk analysis for each hazard discovered and determine actions to take to mitigate or eliminate the risk associated with these hazards.

Hazards include excessive noise, improperly guarded machines, nip hazards, and others as listed in additional topics.

c. To determine a methodology of obtaining the goals.

Possible solutions include:

Periodic inspections by team members of a specific area of the facility. These inspections would rotate through the different areas beginning in the area with the highest accident and incident rates. During these inspections team members at a minimum would verify that proper PPE, stated on the material safety data sheet (MSDS), was in use, that proper lifting techniques were being employed, and that operators whose work involved computer screens and keyboards were using the proper ergonomic techniques.

Establishing training classes by professionals:

For team members at a minimum the administration of first aid, confined spaces, blood-borne pathogens, fall protection, active shooter, diversity, ergonomics, and emergency evacuation routes, and others as required by OSHA.

Training on fall protection must include ladder safety due to the frequency and their costs. Estimates of the cost of ladder injuries in 2011 exceeded 24 billion USD, 113 fatalities, 15,460 on fatal injuries that resulted in lost workdays, 15,460 injuries that caused lost workdays and 34,000 non-fatal injuries that were treated in the emergency department [59].

For all employees at a minimum, the right to know law, fall protection, ergonomics, and emergency evacuation routes and others as required by OSHA.

Insure that a hard copy of a material safety data sheet for every chemical used in the organization exists and is made available to an employee on request. These MSDS files must be an integral part of the Right-to-Know training and will state the PPE that must be properly employed during the use of the specified chemical.

d. To determine which measurements will be used to ascertain progress toward a goal which is the elimination of all accidents and injuries in the workplace and the method of communicating to all employees its current progress toward all goals.

Suggested metrics include the total number of accidents, injuries, and near misses represented graphically, a decomposition of these measurements by department, shift, and cause, and the number of recommendations issued by the safety committee.

e. To insure that after the occurrence of an accident or incident a complete and thorough investigation will be conducted by a knowledgeable person.

f. To insure that each accident, near accident, and incident is properly investigated.

The purpose of investigation is to determine the underlying cause. The prevention strategy often provided is that the employee needs to be more careful. This prevention strategy result will not assist in a further reduction.

Various methods exist, as root cause analysis, which can be used to develop solutions that will prevent future accidents from this cause. The supervisor of the person who was involved in the accident or incident should initiate a complete evaluation in his department to determine if similar hazards or conditions caused this accident to exist elsewhere in his department. The results of this investigation must be conveyed to the safety team so that the existence of the hazard can be determined elsewhere in the organization.

6. Does the company use any hazardous materials in any of its processes?

a. Is the company in compliance with OSHA concerning storage and handling?

Compliance with all requirements must be met and the needed resources provided to insure that these are continually met and exceeded if possible.

b. Are employees properly trained in its usage?

Appropriate training by certified personnel must occur as stated in the regulations and as stated on MSDS sheets provided by the supplier of the material.

c. Do all employees have appropriate PPE?

The MSDS sheet provided by the supplier will indicate the PPE that is required for the storage, handling, and use of the product.

7. Do any operators work in hazardous conditions?

Hazardous conditions include temperature, humidity, ergonomic hazards as continued lifting, confined spaces, biological hazards, physical hazards. A hazardous condition is anything that has the potential to be harmful, cause injury, or have adverse health effects on someone or something [60].

If the operator works in hazardous conditions, the rating factors for those conditions must be added to his normal time when computing standard times.

8. Do any operators have to lift heavy loads as defined by National Institute for Occupational Safety and Health (NIOSH)?

The NIOSH Lifting Equation mobile application, NLE Calc, is a tool to calculate the overall risk index for single and multiple manual lifting tasks. This application provides risk estimates to help evaluate lifting tasks and reduce the incidence of low back injuries in workers [61].

If the operator is required to lift heavy loads as defined by NIOSH, then that element must be retimed using a proper crew and the time normal and standard times adjusted accordingly.

9. How can the working conditions be made more comfortable?

If the operator is required to stand for long periods of time is he supplied with mats on which to stand? Enable employees to perform some tasks be done to the maximum extent possible?

Does the operator have to extend his arms to reach or position in an awkward position? A redesign of the workplace may be needed to prevent accidents, incidents, and near misses and to increase the comfort level of the operator.

Does the operator have to operate any power tools while in an awkward position? This situation must be prevented by a possible redesign of the workstation or the incorporation of robots into the process.

Does the operator have to perform any tasks while on a ladder? This working condition must be eliminated by a redesign of the workstation or the incorporation of robots into the process to prevent possible accidents, incidents, and near misses.

10. What tools and equipment do the operator require to perform his job?

These tools must be properly guarded, the operator fully trained in the use of this equipment and use PPE as required,

The tools must be inspected frequently for the presence of frayed electrical cords, the elimination of the third or grounding been plug removed, the potential overloading of a circuit due to many electrical cords plugged into an outlet, a leak in an air or hydraulic hose, or a pool of hydraulic fluid anywhere around the pump, any hose or the equipment that uses the hydraulic power.

C. CRITICAL QUESTIONS TO ASK AN EXISTING SAFETY PROGRAM WITH SOLUTIONS

1. What are the main successes of the safety team?
   a. The reduction of the rate of accidents and incidents as a result of continuing to investigating the causes of accidents and incidents, determining the causes, and implementing strategies for reduction. Actual results, either positive or negative, must be reported. Positive results indicate a level of success that needs to continue and accelerate. A negative result reveals deficiencies in the program that must be addressed.
   b. The continuation and expansion of training to include additional topics and an increased audience.
   c. To develop a safety incentive program and seek management support.
2. How can these successes be maintained and improved?
   a. By continuing to investigate accidents and incidents and the aggressive implementation of proposed solutions.
   b. By the expansion of existing training programs to include topics as:
      1. The provision of training offered by the American Red Cross which includes basic first aid, cardiac pulmonary suscitation (CPR), and Automatic electronic device (AED) training.
      2. Safety training to prevent back injuries.
      3. Active shooter and diversity training.
      4. A discussion of ergonomics to enable employees to improve productivity and minimize the probability of potential accidents and incidents.
      5. Safety training in the proper use of ladders.
3. Does the company use any hazardous materials in any of its processes?
   a. Ensure that the is currently in compliance with OSHA concerning storage and handling and continues to maintain compliance.
   b. Verify that involved employees are properly trained and that the training is continued and expanded to include other employees as needed,
   c. Appropriate training by certified personnel must occur as stated in the regulations and as stated on MSD sheets provided by the supplier of the material.
4. Do all employees have appropriate PPE?
   The MSD sheet provided by the supplier will indicate the PPE that is required for the storage, handling, and use of the product.
5. Hazardous conditions include temperature, humidity, ergonomic hazards as continued lifting, confined spaces, biological hazards, physical hazards. A hazardous condition is anything that has the potential to be harmful, cause injury, or have adverse health effects on someone or something [60].
   If the operator works in hazardous conditions, then additional rating factors for those conditions must be added to his normal time when computing standard times.
6. Do any operators have to lift heavy loads as defined by NIOSH?

The NIOSH Lifting Equation mobile application, NLE Calc, is a tool to calculate the overall risk index for single and multiple manual lifting tasks. This application provides risk estimates to help evaluate lifting tasks and reduce the incidence of low back injuries in workers [61].

If the operator is required to lift heavy loads as defined by NIOSH, then that element must be retimed using a proper crew and the normal and standard times adjusted accordingly.

7. How can the working conditions be made more comfortable?

If the operator is required to stand for long periods of time is he supplied with mats on which to stand? Optimize the number of tasks that can be performed while the operator is sitting.

Does the operator have to extend his arms to reach or position in an awkward position?

Does the operator have to operate any power tools while in an awkward position? Does the operator have to perform any tasks while on a ladder?

Musculoskeletal MSD injuries were estimated in approximately one million people to lose lost time at work to cause between 45 and 54 billion USD in 2001. The causes of these MSD's are related to work-related risk factors as material handling, frequent movement, rotation of the whole body, and heavy physical work. Changes in various physical aspects in the workplace can drastically reduce the risk for MSD's [62].

All working positions must be evaluated by an ergonomist. Ergonomic training must be a part of the annual safety training that each employee receives.

8. What tools and equipment does the operator require to perform his job?

These tools must be properly guarded and the operator fully trained in the use of this equipment.

The tools must be inspected frequently for the presence of frayed electrical cords, the elimination of the third or grounding been plug removed, the potential overloading of a circuit due to many electrical cords plugged into an outlet, a leak in an air or hydraulic hose, or a pool of hydraulic fluid anywhere around the pump, any hose or the equipment that uses the hydraulic power.

- The location of storage centers is closer to conveyor belts and other work stations.
- The breaking up of heavy shipments into smaller ones to reduce the likelihood of injuries from the lifting of heavy objects or the need for additional assistance in lifting.
- The emphasis of pushing movements over pulling movements, which can strain the lower back.
- The avoidance of twisting motions which increase the likelihood of injury.

- The alternating of tasks among employees provides variation and reduces the risk of boredom and the associated lack of focus on the task.
- The encouragement of employees to wear well-cushioned footwear and the provision of cushioned mats.
- The use of carts or other assistive devices to transport heavy objects.
- Have employees alternate tasks to minimize the burden and allow time for rest.
- Have two or more employees lift heavy objects.

9. Is current training current?

Required training varies by industry and is subject to change [63].

The proper method to store LP tanks which must include visibly marked signage including No Smoking signs that are clearly visible.

- The marking of areas used to recharge electric lift trucks which must be and away from usual vehicular traffic.
- The physical marking of vertical posts and their identification on a drawing to enable locate specifically the position of everyone in case of an emergency.
- Quarterly inspections with local fire inspectors to insure compliance with all fire codes.
  - Included should be opportunities for safety team members to practice using fire extinguishers.
  - Monthly safety inspections with safety hazards identified and corrected immediately with the presence of supervision.
  - Monthly inspections include the chocking of trailer wheels.
  - Random inspections of operators to insure adherence to usage of PPE.

D. EXAMPLES OF SUCCESSFUL AND UNSUCCESSFUL METHODS IMPROVEMENT IN SAFETY PROGRAMS

1. A successful safety team in a furniture plant

The author was a team member of a successful safety team in a furniture company that reduced its accident and incident rate each year. This success was due to management support, monthly training on a variety of subjects, and an active safety team who made inspections weekly to insure that all employees performed their assigned tasks in a safe manner and the used PPE as required, investigated each accident, incident, and near miss. Each investigation listed specific causes which were addressed by the safety team.

The company was recognized by the state for its success in the reduction of workplace accidents and incidents. As a result of the efforts of the safety team, management, and all the employee the costs of workman's compensation insurance annually was reduced from over $150,000 annually to less than a thousand dollars per year.

2. Examples of unsuccessful efforts in safety teams

   a. The author led a safety team of a furniture company that produced unfinished furniture. The team met monthly who met monthly but due to full management support members were not engaged.

   The author unsuccessfully attempted to establish an effective safety to conduct tours to observe operators performing tasks note any potential causes of accidents and injuries to enable the team to make recommendations. The primary reason for the lack of success was a communications issue due to the majority of employees being unable to communicate with all levels of management due to a language barrier.

   This situation should have been addressed before any attempt was made to establish a safety team due to the importance of effective communication. Possible solutions include forming a team with members who were effective communicators with all employees. Another feasible solution would be a requirement that the safety team leader be able to effectively communicate with all employees.

   b. Another example of an unsuccessful attempt at improvement

   The author worked in a textile plant that manufactured narrow fabric with cotton and polyester fiber as raw material. Manufacturing processes included weaving looms and knitting machines. A safety team existed with a supervisor as its leader. Due to the many duties of the supervisor and the high turnover of employees, the team met infrequently and was unable to establish any continuity. Accidents and injuries were discussed during the infrequently held safety meetings with insufficient follow up.

   A potential solution would be the designation of safety team leader who would be responsible to select potential members who would be engaged in developing a culture of safety and obtaining the full support of all levels of management.

   c. Another unsuccessful attempt at improvement in a construction team

   The author was a Contractors Officers Representative (COTR) for a governmental agency for all the contractors who were performing various activities at a governmental medical facility. One of the contracts involved renovations to a section of the hospital to increase capacity.

   Having previous experience with OSHA regulations and as a safety manager, the author stressed the criticality of adherence to the guidelines as well as all existing safety rule. One such guideline was for employees of contractors work to always have a companion.

   One day an employee of a contractor was working alone. One of the tasks in the relocation was to relocate some overhead lights. The contractor had been advised that, although this was a part of the contract, the facility would provide certified electricians at no cost to the contractor if needed to insure that the project was completed on time.

The contractor was advised the facility was to be provided with at least 15 minutes notice before needing the electricians.

The lone employee, who was not a certified electrician, decided that this was a project that he could undertake himself. Using an aluminum ladder he began to remove the light fixtures without disconnecting the electric power that supplied the lights. He would wrap duct tape around the live wires and allow them to hang from the ceiling.

As he was moving the aluminum ladder, the ladder came into one of the live wires hanging from the ceiling resulting in a shortage in the circuit and the activation of the circuit breaker that controlled the power for the entire floor on which the repairs were being made.

Unfortunately, the surgical suites were connected to that circuit. A surgeon was about to initiate a surgical procedure as the lights were denied power. This situation resulted in an unpleasant discussion with the surgeon and the medical director.

As the COTR for this project the author bore ultimate responsibility for the outcome. The author, upon learning of this near miss, contacted the manager of the contracting company and advised him of the situation. The author also advised the manager that the employee who had caused this near miss was to not return to this facility.

The author explained in detail the reason for the expulsion and that the contractor could have been sued for damages if any injuries had occurred to the lack of critical thinking of his employee.

14 Integrating Measurement into a New or Existing Safety Program

A. BRIEF OVERVIEW OF THE NEED FOR MEASUREMENT INTO A SAFETY PROGRAM

The goal of the safety program is a continuous reduction of near misses, incidents, and accidents in the workplace. Recording the volume of accidents and incidents and other relevant information initially and longitudinally will enable the organization to set goals and determine remedial action to achieve these goals.

The goal also includes insuring that all OSHA regulations are met, that recommendations from tours made by the safety team are implemented, continual training occurs, and that the successes of the team and organization are recognized.

B. CRITICAL QUESTIONS TO ASK A NEW TEAM TO INSURE THAT THE MEASUREMENTS INTEGRATED INTO THE INITIAL SAFETY TEAM ACHIEVE RESULTS

1. The initial goal of increasing awareness of current accident and incident rates will be achieved by graphing these two rates and posting the updated graphs on bulletin boards located throughout the company. The graphs will also contain trend lines for both accidents and incidents to demonstrate the results of the safety team. The graphs will also contain a Pareto analysis of the most common causes of accidents and injuries for each period to provide potential improvement opportunities.
2. A graph with the number of violations that the safety team found decomposed by location over time to in awareness of improvements.
3. A graph of the costs of workman's compensation to enable all employees to understand the effect of a reduction in the number and severity of accidents and injuries on these costs. Insure that employees realize that the lower the costs of workman's compensation mean that fewer employees missed work due to accidents and injuries which enables employees to work more hours and increase wages.
4. A graph of the percentage of employees who have received the various types of available and the number who have obtained pay increases to encourage additional training.

DOI: 10.1201/9781003153412-14

C. CRITICAL QUESTIONS TO ASK THE EXISTING SAFETY TEAM

The safety team should focus on those areas that the ratio is not decreasing by an increase in the number of audits performed by the team in those areas, increased training to address the causes of violations, and increased management support.

1. Is the trendline for accidents and injuries decreasing? If not, what actions will the safety team take to reverse the trend?
2. Have the number of violations in the use of hazardous materials or conditions been decreased?

    Each violation should be violated to determine the cause for improvement opportunities. After implementation of improvements, should these improvements be monitored to insure that the desired results were achieved?
3. What percentage of the employees have received the training required?

    Those employees who need training should be trained as soon as possible to meet regulations and as required by their specific jobs.
4. Do all employees have knowledge of the required PPE for this job and use the required for their job and have easy access to that equipment?

D. EXAMPLES OF SUCCESSFUL AND UNSUCCESSFUL TEAMS

1. Examples of successful teams
    a. The success of a team in a furniture plant was due to the use of the accident and incident rates as a measure of progress and continued commitment to educate all employees on the success and the reasons for that success.

        The success was achieved by a combination of the results of safety team weekly audits and the thorough investigation of all accidents and incidents to determine the cause to prevent it from occurring gain, monthly training on a variety of subjects, and an active safety team who made inspections weekly to insure that all employees performed their assigned tasks in a safe manner and the used PPE as required, investigated each accident, incident, and near miss. Each investigation listed specific causes which were addressed by the safety team.

        The company was recognized by the state for its success in the reduction of workplace accidents and incidents. As a result of the efforts of the safety team, management, and all the employees the costs of workman's compensation insurance annually was reduced from over $150,000 annually to less than a thousand dollars per year.
    b. The author worked for a paper converting company with five plants located in the Southeast that manufactured paper tubes that were used for the textile and paper industry. Several plants had identical manufacturing processes.

        The author established a safety team at each plant that was fully supported by all levels of management. Each plant developed a safety checklist that was used during weekly tours to evaluate each process as the operator was performing the tasks needed to manufacture the product. This safety checklist included the use of PPE, the proper guarding of equipment, chocking the wheels of trailers, and the use of chains across the open dock to prevent the possibility of a forklift truck or person from entering an empty loading dock. The lack of restraint across an open dock with no trailer to load resulted in an accident resulting in injuries to the lift driver. Results of each inspection and the corrective actions taken were reported to management.

The results of the team inspections of each plant were totaled and decomposed by department, the purpose was to determine the existence of the same or similar potential hazards for each department across the plants with identical manufacturing processes. This information with the recommendations to reduce the likelihood of injuries and accidents to continually reduce the possibility of potential injuries and accidents was updated and made available to all employees

2. Examples of unsuccessful teams
   a. The author led a safety team of a furniture company that produced unfinished furniture. The team met monthly but due to the lack of full management support members were not engaged.

   The author unsuccessfully attempted to establish an effective safety to conduct tours to observe operators performing tasks to note any potential causes of accidents and injuries to enable the team to make recommendations. The primary reason for the lack of success was a communications issue due to the majority of employees being unable to communicate with all levels of management due to a language barrier.

   This situation should have been addressed before any attempt was made to establish a safety team due to the importance of effective communication. Possible solutions include forming a team with members who were effective communicators with all employees. Another feasible solution would be a requirement that the safety team leader be able to effectively communicate with all employees.

   b. The author worked in a textile plant that manufactured narrow fabric with cotton and polyester fiber as raw material. Manufacturing processes included weaving looms and knitting machines. A safety team existed with a supervisor as its leader. Due to the many duties of the supervisor and the high turnover of employees, the team met infrequently and was unable to establish any continuity. Accidents and injuries were discussed during the infrequently held safety meetings with insufficient follow-up.

   A potential solution would be the designation of a safety team leader who would be responsible to select potential members who would be engaged in developing a culture of safety and obtaining the full support of all levels of management.

15 Integrating Methods Improvement in the Revision of an Existing or the Installation of a New Material Handling Process Change or Packaging Change

A. BRIEF OVERVIEW OF THE NEED FOR MATERIAL HANDLING AND A CONTINUAL REVIEW OF THE EXISTING PROCESS

Material handling is an essential part of any process.

Material must be received, maintained until required for the next process, transferred between processes, and for shipment. Changes in the raw material, the method of receipt, the methods used in packaging of a raw material, the manufacturing processes themselves, or the packaging requirements of the final product for shipment may demand revised methods of receiving and storage of incoming material, the methods used to transport material between processes or a requested revised method of packaging required by the customer could result in a study to investigate changes in packaging.

B. CRITICAL QUESTIONS TO ASK FOR THE INVESTIGATION OF A NEW METHOD OF MATERIAL HANDLING OR SHIPMENT TO THE CUSTOMER WITH SOLUTIONS

1. Is this new packaging designed for an existing product?

    This will require that the flow of an existing product involved in this new packaging method be evaluated for the potential affects. If affected, then the method of transporting existing product in between processes must be evaluated to determine necessary changes. For example, if the current method of packing is into a carton of 12 each and the proposed method consists of a unitized package containing two rows of 12 each stacked onto each other before the package is wrapped, what material handling modifications will be needed to be made to accommodate the larger and heavier package?

DOI: 10.1201/9781003153412-15

The changes must be included in the ECO requesting the revision. The costs of these changes will offset the savings that would accrue from the packaging change. The results of any testing the new package performed must be attached to the ECO.

2. Is this a new package for a new product?

The method of packaging, the results of the test, the equipment with the costs of installation, and the packaging cost per unit must be included in the ECO requesting this method of packaging.

This information is needed to calculate the packaging cost per sales unit. If the selling price is based on each or per item, then all costs of manufacturing must be determined per unit.

C. CRITICAL QUESTIONS TO ASK FOR MODIFYING A MATERIAL HANDLING SYSTEM OR CHANGING METHODS OF PACKAGING

1. Why is the request to modify existing packaging being made?

If this is a request by a vendor, then several samples of product that meet the requested changes must be supplied to insure that the new method meets the needs of the customer. If this new packaging uses recyclable materials, are those materials easily identifiable and the procedures established for their recycling?

If there are costs increase due to this requested change, then a negotiation process should be initiated in an attempt to recover some of these increases.

2. Is this request an internal one? If so, then the proposed method must be simulated internally to insure that this new method is feasible and costs determined to validate that a cost reduction will occur either in the packaging process or another process. The new method of packaging must be visually approved by the customer. The customer needs input from receiving and manufacturing before accepting the change. Then a sample of the product with the new proposed packaging must be shipped to the customer for evaluation. Depending on the amount of savings and the payback associated with the change, the vendor may offer a discount to the customer for accepting the revised packaging.

3. What is the reason for the request?

There are numerous reasons to modify existing material handling systems.

a. To reduce material handling costs.

Material handling begins upon initial receipt and includes the transfer of material as it is processed and continues until the product is shipped.

A value stream map or flow diagram for each product will reveal the occurrence, frequency, and duration of each material handling operation. A work sampling study can be employed to determine the duration of each occurrence of a material handling process. This time, converted to labor hours per sales unit, and the frequency of each duration is used to calculate the indirect labor costs per sales unit. This information, indirect labor cost per unit, multiplied by the quantity of sales unit affected by the proposed material handling change is the indirect labor costs for that particular, operation.

The total indirect labor cost for each product would be the sum of the indirect labor costs of all the indirect labor costs. This total cost would provide input for the costs justification for the proposed material handling system.

It must be emphasized that the indirect labor savings as a result of the new material handling system are true savings only if the indirect labor used prior to the installation is transferred and fully engaged in the new position. If that does not occur, then in actuality it is that the money spent for the new system was not fully justified and that additional unforced idle time is provided for indirect labor.

b. To reduce in-process waste.

Often the unintended consequence of a material handling system is the generation of defective products. A process flow diagram or a value stream map will provide those material handling processes in which this occurs and the amount or percentage of defective products produced. These defective products represent lost sales dollars. These lost sales dollars are a metric that can be used to justify the costs of the installation of a new material handling system.

c. To reduce the floor space required for processing.

In addition to the installation of material handling equipment, another manufacturing method to reduce footprints needed for the manufacturing process is to incorporate cellular manufacturing.

The use of cellular manufacturing can also reduce direct labor costs, work-in-process inventory, and increase throughput.

d. To reduce the probability of an accident, near miss, or injury.

Ergonomics must be considered as new material handling systems are designed that will eliminate motions as twisting, lifting heavy objects. The reduction of hazardous materials and conditions due to the new material handling systems.

D. EXAMPLES OF SUCCESSFUL AND UNSUCCESSFUL CHANGES IN MATERIAL HANDLING AND PACKAGING

1. Examples of successful changes
   a. An example of a packaging change that initially was unsuccessful but was successful after the use of critical thinking and input from others.

A paper tube, think of a straw, was supplied to manufacturers of cloth, denim, material for T-shirts, and similar products. These tubes, only were manually removed from the conveyor after being heated, placed into a fixture, manually tied into a bundle of 19, 24, or 37 tubes per bundle, depending on customer specifications using a fixture, and stacked onto a pallet for subsequent handloading into a trailer. This was a very labor-intensive process that was repeated after the trailer was received at the customer's dock. The various material produced by the customer would then be wrapped around the paper tube for shipment.

This figure was initially presented in Chapter 9 (Figure 15.1).

This figure was initially presented in Chapter 9 (Figure 15.2).

The author envisioned a unitized package that could be loaded onto a trailer with a forklift. This new packaging method would reduce the time required to load the trailer. The process of removing tubes from the conveyor, placing them into a fixture with the correct quantity, tying the bundle together with three bands, and positioning the tied bundle onto a pallet was eliminated. The new process would require the loading of the tubes into a fixture, operating the stretch wrap machine to unitize the package, lifting of the unitized package out of the fixture via a forklift, and loading into a trailer for shipment. Upon arrival at the customer, the trailer would be unloaded via forklift.

This proposed method of packaging would eliminate the process of placing the tubes into a bundling rack, tying the bundles, placing the tied bundles onto a pallet and manually unloading the pallet, and

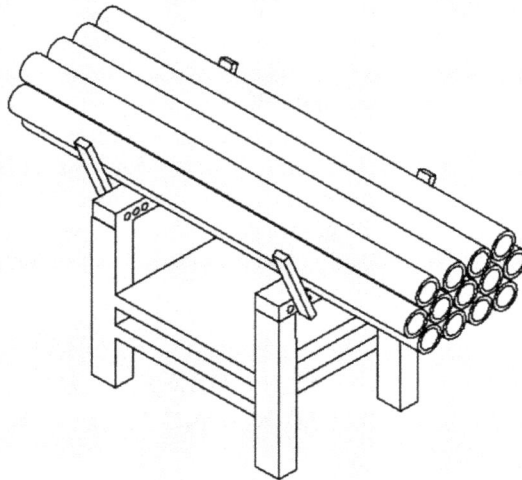

FIGURE 15.1 Fixture for racking tubes manually (orthogonal view).

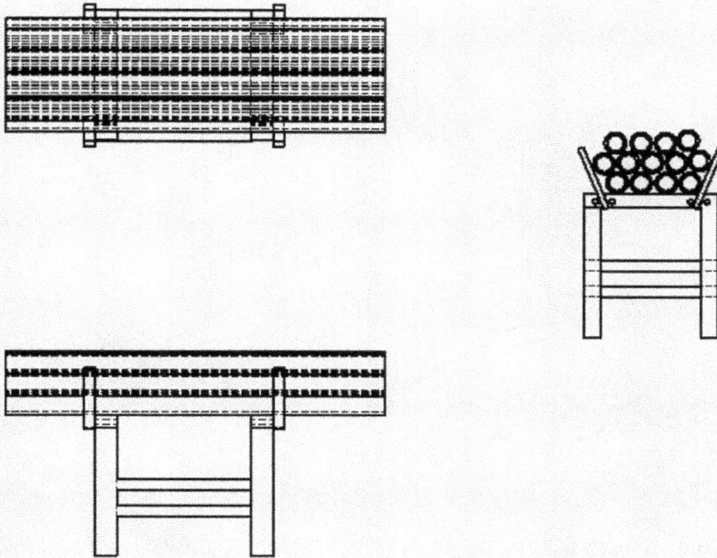

FIGURE 15.2 Fixture for racking tubes manually (plan, top, and right-hand side views).

loading the truck. It would also eliminate the reversal of these tasks after arrival at the customer since unloading would be using a forklift.

The next step was to discuss the possible change with management. Management agreed to the concept. A fixture was fabricated and a sample roll of stretch film was acquired to process a sample pallet (Figures 15.3 and 15.4).

The author discussed the change with the plant manager who was in full agreement with attempting this new method. An inexpensive stretch wrap machine was purchased and delivered as well as several rolls of the most expensive stretch film available – after all the film was samples and free. With the help of maintenance, a fixture was designed and built (Figures 15.5 and 15.6).

The author discussed the proposed packaging change with a salesman to obtain a potential customer who would accept the change. The customer explained the potential change with receiving and manufacturing to insure that this new packaging method would not result in additional material handling concerns which would increase indirect labor costs. Both functions agreed to a trial order and indicated that the new method would reduce the time consumed in unloading the trailer and resupplying each machine. On the next order that customer, the fixture was placed adjacent to the oven to enable the employee to load the

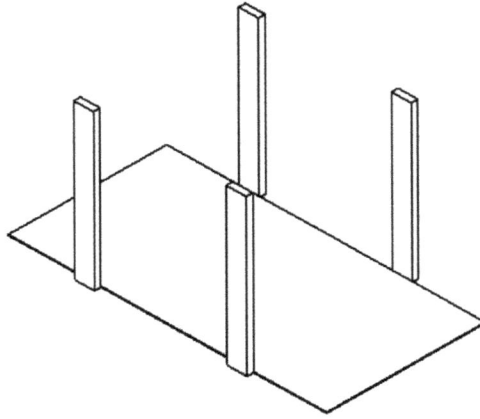

FIGURE 15.3 Empty fixture into which tubes are manually placed against a backstop to insure tube alignment and to verify the correct length (orthogonal view).

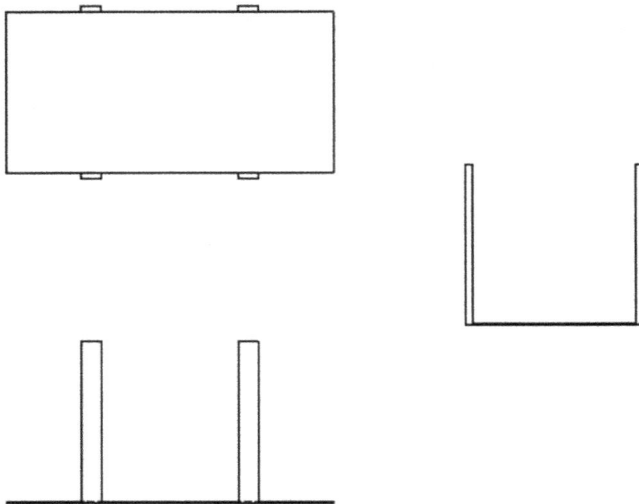

FIGURE 15.4 Empty fixture into which tubes are manually placed against a backstop to insure tube alignment and to verify the correct length (top, plan, and right side view).

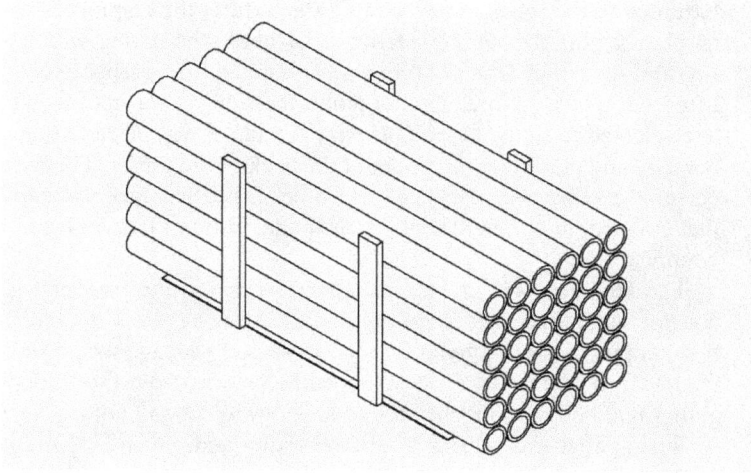

FIGURE 15.5   Filled fixture with tubes ready to be wrapped – notice the vertical alignment of the tubes as a result of the use of the backstop (top, plan, and right side view).

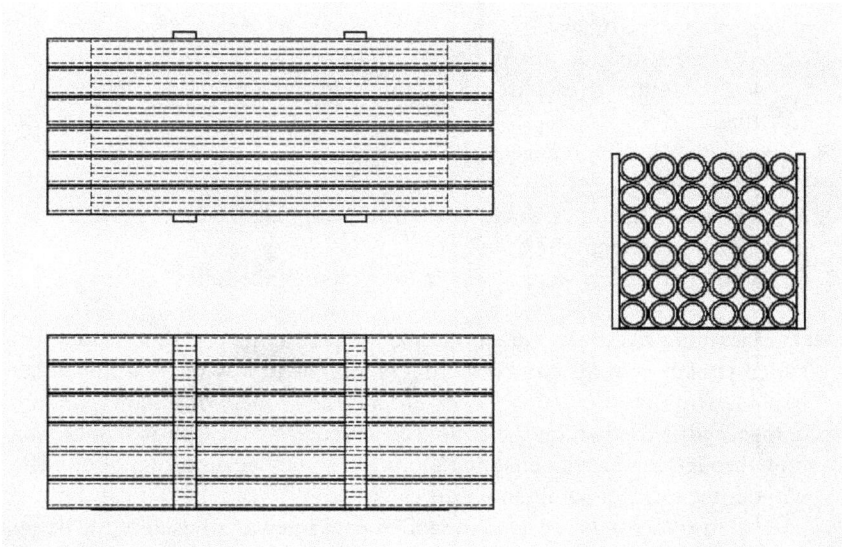

FIGURE 15.6   Filled fixture with tubes ready to be wrapped – notice the vertical alignment of the tubes as a result of the use of the backstop (top, plan, and right side view).

fixture with the tubes. After loading the fixture, the employees placed the fixture onto the stretch machine adjacent to the fixture with a fork-lift, and the pallet with the tubes was wrapped, the wrapped load was lifted out of the fixture. At this point, the tubes were hot having just been ejected from the oven. This wrapped pallet was placed aside tem-porarily and placed on the trailer at the back of the trailer. The remain-der of the order was packaged as before. The customer was advised that the next order would contain one pallet of tubes that was stretched wrapped.

The engineer was at the customer to observe the wrapped pallet and for input from the receiving and material handlers that supplied the equipment. Both agreed that the proposed package would be ben-eficial. Upon removal of the wrapped pallet was from the trailer, the author realized that the once round tubes were flat and not usable. The wrapped pallet was reloaded onto the trailer and returned to the plant for recycling.

Upon returning to the plant, the author explained the situation to various levels of management to obtain their input as to the cause and their suggestions for improvement. Management suggested that the tubes after being wrapped were not able to cool and as a result retained the moisture received from the application of the adhesive and the weight of the tubes caused the tubes to collapse. This was verified by comparing the tubes at the lower levels of the pallet with those higher in the pallet. The flatter tubes were at the lower levels; the lower the level the flatter the tubes.

The reason that this initial attempt to revise the packaging method was the failure to employ critical thinking and not seek assistance from others.
The questions that were not asked:

1. What effect on the tubes would result from wrapping the tubes immediately as they emerged from the oven?
2. What other materials were available for wrapping the tubes?

The author researched other methods of packaging and found a product called stretch netting. Stretch netting is similar to a wire mesh fence but is made from plastic and stretches around the corners of a pallet without damaging the product on the pallet. Its advantage over film is that air can flow through the netting enabling a product to be packaged out of an oven without the product being damaged.

The customer who received the failed package was advised of the better material for wrapping the tubes that would enable the tubes to dry after the pallet was wrapped and asked if another sample with this material could be shipped for evaluation. On the next order for this customer, a pallet of tubes was packaged with the stretch netting, loaded onto the back of the trailer, and was observed by the author, the PA and supervisor from receiving and manufacturing. Inspection of the tubes after the pallet was unloaded revealed that this method was successful; all tubes in the pallet that had been wrapped with the netting were usable.

The author returned to the plant and advised management. Management suggested that two pallets be wrapped and stacked to simulate the loading of the trailer. The results indicated that two pallets could be stacked with no deterioration of tube quality. These two stacked pallets remained in the plant for random observations for a week. To simulate the effects of this new packaging on tubes that remained in the trailer overnight to simulate shipping to distant customers, two pallets by placing two new wrapped pallets into the nose of a closed trailer to remain overnight. Inspection revealed that there was no deterioration in any of the tubes, particularly those in the bottom rows.

The customer was advised that additional testing with positive results had occurred and requested permission to initiate shipping using the new method. The customer agreed. The author was at the loading dock of the customer when the next order arrived. These pallets were inspected and approved by the author, the PA, and the supervisor of receiving and management.

This proposed change initially failed because the author did not fully research alternatives, receive samples of each alternative, and obtain feedback from more experienced personnel. The author sacrificed an initial success for time. As Thomas Jefferson said, "Take time for all things. Great haste makes great waste."

b. A successful change in material handling

One plant that manufactured paper tubes for the textile industry packed the tubes in corrugated containers. These cartons were then stacked three high, transported manually on a hand truck into a trailer, removed from the hand truck, and then placed as needed in the trailer. The cartons were handled singly due to the weight and size of the loaded carton.

An attempt that was made to capture the amount of time needed to load a trailer manually was not successful due to the intermittent occurrence of this task. Cartons were transported to the trailer only after three cartons had been filled and often waited on the material handler since he had other tasks to perform. The author met the fully loaded trailer at the customer's dock to observe the unloading process.

Upon opening the trailer doors at the customer's dock, a forklift with carton attachments began unloading the trailer, 12 cartons at a time. The lift truck moved the cartons into the receiving department and returned to remove another 12 cartons. This process was repeated until all the cartons on the trailer had been taken to the receiving department. The author timed the unloading operation which required eight minutes.

Upon returning to the plant, the author investigated the use of carton clamps to load the trailer. Management was initially skeptical but when advised that a clamp was used to unload the cartons and place them into receiving and to a machine as needed but realized that if the trailer can be unloaded with a clamp, then it could be loaded with a clamp.

The lift truck dealer that supplied the lift trucks in the plant was called and brought an adjustable clamp that could be used by one of the lift trucks in the plant. This lift truck with the attached clamps began to immediately pick up cartons for loading. Tubes with cartons loaded with the clamps were examined for distortion. The result of these examinations concluded that the use of carton clamps could successfully be used to reduce the time required to load a trailer and movement between processes. Attachments are also available that can be used for rolls of material.

The author and management immediately adopted this new material handling method. Although the cost savings could not be determined due to the inability to accurately measure the indirect labor hours needed to load the trailer due to the intermittent loading process, management began to employ this method due to the elimination of the manual lifting required by the previous process.

2. An unsuccessful material handling change

One aspect of the job was that the author enjoyed traveling with salesmen to customers to assist in solving problems or providing technical assistance. The company served a large manufacturer of vinyl-backed wallpaper. The current method of packaging was all manual and consisted of several steps.

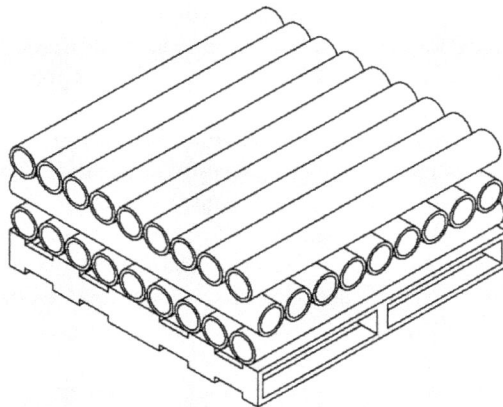

FIGURE 15.7 Pallet illustrating "pig-penned" or the packing tubes in rows that are perpendicular to each other for package integrity (orthogonal view).

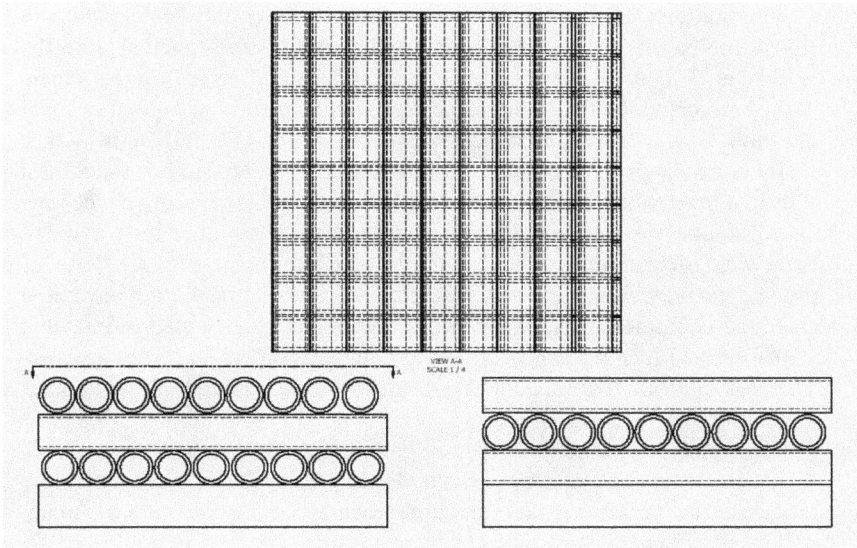

FIGURE 15.8 Pallet illustrating "pig-penned" or the packing tubes in rows that are perpendicular to each other for package integrity (top, plan, and right side view).

After the roll was produced and wound onto the paper tube, the roll was placed onto a conveyor belt where a crew of two first removed a predetermined length of brown kraft paper and then wrapped the roll the brown kraft paper, and secured it with masking tape. The rolls were then manually placed onto a pallet where each additional layer was placed perpendicular to the layer below. This arrangement is usually referred to as the tubes that were 'pig-penned' onto the pallet for an unsuccessful packaging change (Figures 15.7 and 15.8).

The pallets were then wrapped and strapped onto the pallet using a polypropylene strap which is safer and less expensive than steel strap but more expensive than plastic strapping which at the time had a tendency to stretch.

The plant engineer explained that the current method of packing was labor intensive and therefore expensive and his proposed method was to be implemented the following Monday after the equipment was installed over the weekend. This new packaging method consisted of the use of a robot to obtain the finished roll, hold it in position as a horizontal stretch wrap automatically wraps the finished roll, and then places the roll onto a pallet in a pig-penned style. A crew of two would then place a shrink over the entire pallet place and then shrink the bag using handheld devices using propane as fuel. The pallet would be available to be then loaded into a trailer with a forklift. He advised the calculated payback was less than six months due to the costs of manually wrapping the rolls.

The author's concern focused on whether or not the heat applied to shrink the bag would result in causing the vinyl to fuse together with the wallpaper. When the plant engineer said that he had asked the engineers with the companies that manufactured the heat shrinking equipment that question, he was assured that the heat would have no effect, then the author really began to get concerned for a fellow engineer. The author asked him if he had tried this process out personally before recommending the change and of course his response was that he had not. The author then asked if there was a hairdryer around that could be used for reassurance that this process would work as planned. His response was that he did not have time since the equipment was on site, work orders had been signed to installed it that weekend, crews and been trained, and customers and been advised. The only question that was not asked was if he had time to look for another job before the weekend.

The following Monday the author called to talk with the engineer over the project only to discover he was no longer with the company. This mistake cost this engineer his job, his reputation and not to mention the many dollars the company lost as a result of the lack of critical thinking. His desire to save the company money and improve processes was the purpose of his position but he allowed the desire to demonstrate results quickly to overcome proven tried and true methods for process improvement. First, ensure that the process works for your product, second, check to determine if other companies who produce similar products use this process, three, document both your successes and failures and the reasons for both, and four, proceed in a logical methodical manner to ensure that the results are meaningful and reproducible.

Questions that should have been asked but were not resulted in failure.

1. Why did the engineer not test the effect of heat on the vinyl-backed wallpaper by using a handheld device as a hairdryer to apply the heat needed to shrink the bag over a roll of wallpaper?

    This test would have required a few minutes to perform and demonstrated that this method of packaging would not work as proposed.

2. Were any manufacturers of vinyl-backed wallpaper using this method of packaging?

    If other manufacturers were employing this or a variation of this method, this information was available from suppliers of the equipment needed or from the customers this company supplied.

16 Integrating Measurement in Installing a New Material Handling System or Packaging Method or the Revision of an Existing Material Handling System or Packaging Method

A. BRIEF OVERVIEW OF THE NEED TO INCORPORATE MEASUREMENT IN THE REVISION OF AN EXISTING OR THE INSTALLATION OF A NEW MATERIAL HANDLING CHANGE

The goal of every organization must be continuous in all aspects of the organization. The installation of new or the modification of an existing material handling system can result in a decrease of direct and indirect labor costs, material costs as a result of a reduction in defects, near misses, accidents and injuries, raw material and work-in-process inventory levels, and profitability.

Increasing profitability requires that all projects be cost effective.

B. CRITICAL QUESTIONS TO ASK WHEN INSTALLING A NEW MATERIAL HANDLING OR PACKAGING METHOD

1. Why is the change being made?

The change may result in a cost savings of direct labor or a revision in the packaging requirements.

The current costs, including direct labor and material must be computed and compared with the costs after implementation of the new method. This cost must include any costs of utilities, the total costs of installation which

DOI: 10.1201/9781003153412-16

could include interest or lease payments depending the method used to obtain the equipment.

2. Will indirect labor be affected?

   This change could be the result of additional or a reduction in the crew size. Any changes need to be verified by an industrial engineer and documented in process routings, the process sheet, the bill of material, and other affected documentation

3. Will additional training be required?

   Additional training may be needed for the operators and quality inspectors

4. Who will initiate the ECO and assure that all documentation affected by this change is included?

   The person responsible must include all those affected by the change and will include at a minimum purchasing, engineering, plant management, quality and the material planner who will establish the effectivity date.

5. Will the customer be affected by this change?

   The customer must be included in the decision to change packaging due to potential effects on receiving and internal Material handling. If the material handling changes that will involve only the internal manufacturing facility, the customer does not need to be advised.

C. CRITICAL QUESTIONS TO ASK WHEN REVISING AN EXISTING MATERIAL HANDLING OR PACKAGING SYSTEM

1. Why is the change being made?

   The change may be required due a change in the unit of measure purchased, a request from the customer, a change in specifications, a process redesign, the need to reduce the costs of direct labor, or a revision in the packaging requirements.

   The current costs, including direct labor and material must be computed and compared with the costs after implementation of the new method. This costs must include any costs of utilities, additional costs of needed training for operators and inspectors, the total costs of installation which could include interest or lease payments depending the method used to obtain the equipment.

2. Will indirect labor be affected?

   This change could be the result of additional or a reduction in the crew size. Any changes need to be verified by an industrial engineer and documented in process routings, the process sheet, the bill of material, and other affected documentation

3. Who will assure the accuracy of any documentation affected?

   The industrial engineer will be responsible for the accuracy of the bill of material, purchasing for costs, the material planner for the effectivity dates, and inventory management for the accuracy of the inventory.

4. Who will initiate the ECO and insure that all documentation affected by this change is included?

The person responsible must include all those affected by the change and will include at a minimum purchasing, Engineering, plant management, quality and the material planner who will establish the effectivity date.

5. Will the customer be affected by this change?

The customer must be included in the decision to change packaging due to potential effects on receiving and internal Material handling. If the material handling changes that will involve only the internal manufacturing facility, the customer does not need to be advised.

D.   EXAMPLES OF A SUCCESSFUL AND AN UNSUCCESSFUL MATERIAL HANDLING AND PACKAGING CHANGES

1. Example of a successful change

The example of the successful packaging change that involved the conversion of manual loading of the unitized package using stretch webbing was successful due to the use of critical thinking to generate a list to be taken and questions to be answered.

Measurements to be taken.

a. The number of indirect labor hours needed to transfer the loaded pallet into a trailer, the time needed to remove the bundles of tied tubes into the trailer, and the time needed to remove the empty pallet from the truck for reuse. A formal time study was used to determine this measurement. The time needed to load the fixture used to tying the bundle and the placement of the tied bundles of tubes onto a pallet for loading into a trailer for shipment was not needed to determine costs savings from this project. After conversion to the new method, the bundler's task would be the loading of the fixture atop the stretch wrap machine and properly applying the stretch netting to secure the load of tubes, using the available fork truck to remove the wrapped pallet from the stretch wrap machine into the truck.

b. The amount of stretch netting used to wrap each pallet. Measurements of the amount of netting used on ten pallets were taken to determine an average. Using the costs of netting per linear feet was computed and converted into costs per M tubes. The dimensions of the packaged tubes varied directly with the length of the tubes, the quantity per pallet varied inversely with the outside diameter (OD) of the tube.

c. The amount of twine used to bundle the tubes. Each bundle was tied three times but the amount of twine used was directly related to the OD of the tube.

These measurements were taken to determine the costs of stretch wrapping per M tubes versus the cost of the twine. Production of tubes produced in a recent month was used to calculate the costs of each per M tube. The results revealed that the additional costs of the use of stretch net were negligible and thus were not considered in the project evaluation. The primary reason for the negligible was that although the costs per linear foot were greater than the cost of twine, there was a sufficient difference in the amount of usage to offset the costs difference.

Savings from this project resulted from the elimination of the indirect labor to load the trailer and from a reduction in the amount of time required to unload the trailer. This reduction in the time to unload the trailer enabled the increase in usage of the trailer.

The total savings would be the product of the number of indirect labor hours saved and the cost of an indirect labor hour.

These calculated savings are realistic if the indirect labor is transferred to other duties and is fully engaged in these duties. A work sampling study needs to be conducted to insure that the indirect labor that was required to load the trailer is fully engaged in other duties; if not the result of the conversion is an increase in the amount of unforced idle time.

The information needed to evaluate the change in the process of loading cartons into a trailer was the time needed to load the trailer using the lift truck with a carton clamp attachment, the amount of time required to load a trailer manually, and the amount of defective tubes that resulted from the use of the clamp.

A formal time study was used to measure the time to load a trailer using the lift truck with the carton clamps. The attempt to determine the loading time of a trailer was unsuccessful due to this being a stochastic process and the time for each trip varied due to variability in the starting location. Observation of tubes that had been subjected to the clamp resulted in concluding that the use of the clamps did not increase the number of defects.

Thus net savings resulted from the reduction of indirect labor hours needed to manually load the trailer. To dollarize these savings would require the number of hours saved to be multiplied by the costs of an indirect labor hour,

2. Example of an unsuccessful change

The proposed change involved replacing manual wrapping of vinyl-backed wallpaper with brown draft paper with automatic wrapping with stretch film.

The reason for the failure was that the heat applied to the shrink bag resulted in the adherence of the shrink to the stretch film.

The following measurements were taken by the engineer to justify the attempted process change.

1. The time to wrap each roll of wallpaper with brown kraft paper. A formal time study was used to determine this information. Included in the time was unrolling the paper and cutting it the desired length, wrapping each roll with the paper including each and securing the roll with a predetermined length of two-inch-wide packaging tape, and adding the roll into position onto the adjacent pallet. This time was converted into direct labor hours per

roll. Crew size was two. The product of the direct labor hours per roll and the direct labor costs yields the direct labor costs per roll for wrapping the roll with kraft paper.

2. The amount of brown paper used to wrap each roll was measured in linear feet. This was converted to cost per roll using the most recent invoice after converting the purchasing unit of measure to linear feet if needed.

3. The quantity in linear feet of packaging tape per roll. This was converted to costs per wrapped roll by obtaining the product of the quantity per roll and the cost per linear foot that is retrieved from the most recent invoice.

4. The amount of stretch film applied to the roll. This was converted to costs per wrapped roll by obtaining the product of the quantity per roll and the cost per linear foot that is retrieved from the most recent invoice.

5. The length of time required to apply the stretch bag over the pallet and to shrink the bag. A formal time study was taken to the quantity and converted to direct hours per roll. Crew size was two. This time was converted into direct labor hours per roll. The product of the direct labor hours per roll and the direct labor costs yields the direct labor costs per roll for the application of the shrink bag.

6. The length of time needed to apply banding to a loaded pallet. A formal time study was taken to the quantity and converted to direct hours per roll. Crew size was two. This time was converted into direct labor hours per roll. The product of the direct labor hours per roll and the direct labor costs yields the direct labor costs per roll for the roll application of banding.

7. The quantity of webbing per pallet and the cost per linear foot from the most recent invoice divided by the number of rolls per pallet gives the webbing cost per roll.

The cost of packaging with the existing method is the sum of:

1. The direct labor costs of wrapping the roll with kraft paper.
2. The material costs of the kraft paper.
3. The material costs of the packaging tape used to secure the kraft paper onto the roll.
4. The direct labor costs of applying and shrinking the bag.
5. The cost of the shrink bag.
6. The cost of the propane used to shrink the bags.

The sum of these costs converted to costs per roll is the packaging costs per roll of vinyl backed with the current method of packaging.

The costs of packaging with the proposed method would include:

1. The material costs of the stretch film applied to each roll.
2. The direct labor costs of applying and shrinking the bag that was placed over the pallet.

3. The cost of the shrink bag.
4. The cost of the propane used to shrink the bags.

The sum of these costs converted to costs per roll is the packaging costs per roll of vinyl backed with the proposed method of packaging.

The savings from conversion to the new method is the difference of the two costs. These savings result from the elimination of the manual process of wrapping the roll in kraft paper and the difference in costs between the kraft paper, the shrink wrap, and the vinyl used for the backing of the wallpaper.

APPENDIX
Scientific Management

Scientific management is a recent managerial version of Human Resource Management (HRM) to organize a group of individuals to attain one or more specific goals. These various methods have existed since the appearance of humans on earth. A Chinese philosopher, Menius, who lived from 372 to 298 BC promoted the various techniques of production management and the division of labor. For example, during the ancient Greek civilization the concept of uniform work methods was understood and practiced. Charles Babbage in his 1932 publication, "On the Economy of Machinery and Manufacturers", promoted decomposing each job into specific elements and determining the cost of element individually. This would result in the ability to calculate individual savings from the adoption of various changes. The concepts of "shop management" and "shop accounting" were proposed to the American Society of Mechanical Engineers by Henry Towne who lived from 1844 to 1924 [64]. Thus human resource management has morphed and experienced growth over time.

The father of Scientific Management, Frederick Taylor, began his career with Midvale Steel after graduation from the Stevens Institute of Technology. In his 1909 publication of "The principles of Scientific Management", he proposed four principles which were focused on determining the most efficient method of performing each job using analysis, the selection of the best person for the job, tracking output and providing training as needed, and to divide the work among management and labor. His use and analysis of time studies resulted in the redesign of shovels used in the loading of coal to feed ovens in the Bethlehem Steel Mills that reduced the number of employees performing this function, 72%, from 500 to 140.

Implementation of these principles occurred in many manufacturing facilities, often tripling productivity. These principles were also applied in household tasks based on the results of time and motion studies. Although application of these scientific management principles benefited from the improved productivity and a significant positive effect on industry, it came with drawbacks as the increased the monotony of work and reduction in the variety of skills, tasks, significance, the autonomy, and the feedback among employees and management [65].

Frank and Lillian Gilbreath were other pioneers in scientific management who made overlapping contributions to the field. Time studies and optimum work methods were the primary focus of Taylor. He was more interested in increasing the output per person by focusing on the selection of the right person for the job and training, whereas the Gilbreaths were more interested in the relationship between motion, movement, and productivity to eliminate unnecessary activity in order to determine to best work methods to increase productivity.

Frank began his career as a masonry contractor. His major contribution to Scientific Management and Industrial Engineering was the application of motion analysis to the workforce. He developed adjustable scaffolding to keep his masons on the same level as the wall as the wall they were building, made improvements to existing cement mixers and methods to drive pilings quicker. All of these modifications resulted from his H of masons while working and realizing that each mason was wasting motion and his goal was to eliminate the wasted motion and energy. One example occurred in the task of bricklaying that reduced the time from 18 to 4 ½ sec or a 75% reduction in the time required to lay one brick. This was achieved through the invention of an easily adjustable scaffold and a method of stacking the bricks on the scaffold to enable the bricklayer to obtain the brick with one hand and apply mortar with the other hand.

He studied an operation and with the use of film was able to decompose each human motion into 17 elemental motions, referred to as therbligs, Gilbreath spelled backwards with modifications. Therbligs was a method of decomposing the motions involved in the accomplishment of a task. The system was designed to locate and eliminate unneeded motions that resulted in the waste of time. This system was invented and improved on between the years 1908 and 1924 [66].

Lillian Gilbreath, the mother of modern management and Frank's wife, worked with him in his consulting firm. Their stated purpose of motion study was to reduce and eliminate unnecessary fatigue, which could be accomplished by the design and placement of workbenches and chairs located for the employees that could be used during regular rest periods. Dr. Gilbreath also believed that job satisfaction and indirect incentives as money were employee motivators. Together they worked to improved job and work simplification, job standardization, job simplification, a wage incentive system, and improved methods to increase employee satisfaction. Recognition of the effects of fatigue and stress on time management was a priority of Lillian and studies were made to mitigate that effect. She was the first female to become a member of the American Society of Mechanical Engineers.

Dr. Gilbreath, the mother of 12 children, studied Scientific Management with her husband but with a PhD in Psychology, she was more interested in the aspects the workplace due to her keen empathy for people as well as her insight into human behavior. As a consultant she worked for companies applying psychology to address problems in the workplace and in the home. Some of her designs to simplify everyday life and make necessary work easier included the shelves inside the refrigerator and the foot pedal activated trash can. She and her husband were the parents of 12 children and she was the author of the book *Cheaper by the Dozen*. In addition to her consulting work she taught college at several schools, including Newark College of Engineering, Bryn Mawr, Rutgers, and the University of Wisconsin. At MIT she was appointed a resident lecturer [67].

The Gilbreaths were also interested in increasing productivity but their focus was in a different direction. They would film an operation to enable them to study the activity in detail. Since they were devoted to improving employee welfare and motivation, they would decompose the operation into discrete elements so that could possibly reduce the motions needed for an operation. The result of such an elemental breakdown was that it enabled a timekeeper to determine a time for that task and the element of which it was a component to be taken for comparison purposes to track changes in productivity against changes made on the production floor.

The result was that much of the work of their research was incorporated in quality control and assurance programs that were initiated during the 1920s. Over time the study of ergonomics resulted from their cumulative work.

Lillian Gilbreath, the mother of modern management and Frank's wife, worked with him in his consulting firm. Their stated purpose of motion study was to reduce and eliminate unnecessary fatigue which could be accomplished by the design and placement of workbenches and chairs located for the employee's that could be used during regular rest periods. Dr. Gilbreath also believed that job satisfaction and indirect incentives as money were employee motivators. Together they worked to improved job and work job standardization, job simplification, a wage incentive system and improved methods to increase employee satisfaction. Recognition of the effects the effects of fatigue and stress on time management was a priority of Lillian and studies were made to mitigate that effect. She was the first female to become a member of the American Society of Mechanical Engineers.

Dr. Lillian Gilbreth contributions include the application of psychology to employees to improve the satisfaction level of employees realizing that an increase in satisfaction would result in an increase in job satisfaction and productivity. A focus of Dr. Gilbreth was the physical comfort of the employee that eventually resulted in the field of ergonomics.

Dr. Gilbreath, the mother of 12 children, and author of the book *Cheaper by the Dozen* was more interested in the mental and physical aspects of the workplace due her keen empathy for people as well as her insight into human behavior. As a consultant, she worked for companies applying psychology to address problems in the workplace and in the home. Some of her designs to simply everyday life and make necessary work easier including the shelfs inside the refrigerator and the foot pedal activated trash can [68].

Another pioneer in Scientific Management was Henry Gantt, who worked alongside Frederick Taylor for a number of years. He is best known for the development of the chart that bears his name. This chart is used in project management and a graphic representation listing in sequential order the tasks that comprise a project. The chart depicts starting and ending points for each task that must be completed in the project. Times for completion are also provided to enable the manager to maintain control over the project.

The other development, probably not as well known, is the task and bonus plan which is a modification of the "fair day's pay for a fair day's work or task". The purpose of this program was increase an employee's pay after an employee attained a certain level of productivity.

Another development is the progressive pay system whose goal was to pay employees who desire to produce additional output for additional income [69].

Bibliography

1. Some Famous Unit Conversion Errors. (n.d.). Retrieved 1/23/2021 from https://spacema th.gsfc.nasa.gov/weekly/6Page53.pdf#:~:text=Some%20Famous%20Unit%20Conversi on%20Errors%21%2053%20Story%201%3A,off%20course%20-%20-%2060%20mi les%20in%20all
2. The Official Definition of Industrial Engineering as Defined by the Institute of Industrial and Systems Engineering. Retrieved 2/5/2021 from https://www.iise.org/ details.aspx?id=282
3. Gilbreth, Lillian Moller. (1872–1972). Lillian and Frank Gilbreth—The Birth of Ergonomics. Retrieved 11/15/2019 from http://www.theinventors.org/library/inventors/ blGilbreth.htm#:~:text=A%20pioneer%20in%20ergonomics%2C%20Gilbreth%20pa tented%20many%20kitchen,Studies%2C%20which%20supported%20work%20simpli fication%20and%20industrial%20efficienc
4. Mee, John F. (2021). Frederick W.Taylor—American Inventor and Engineer. Retrieved 11/15/2019 from https://www.britannica.com/biography/Frederick-W-Taylor
5. Nielson, Jacob. (3 Jan 2012). Usability 101: Introduction to Usability. NN/g Nielsen Norman Group. Retrieved 6/28/2019 from https://www.nngroup.com/articles/usability-101-introduction-to-usability
6. The Saylor Foundation. (n.d.). Scientific Management Theory and the Ford Motor Company. Retrieved 3/17/2021 from https://resources.saylor.org/wwwresources/arc hived/site/wp-content/uploads/2013/08/Saylor.orgs-Scientific-Management-Theory-a nd-the-Ford-Motor-Company.pdf
7. Pepin, Shayna. (2012). How Industrial Engineers Will Save Health Care. *Health & Medicine, Industrial Engineering*, XV(lll, Dec 7). Retrieved 1/16/2020 from https://il lumin.usc.edu/how-industrial-engineers-will-save-health-care/
8. Watson, Gerald J. (2009). Using Human Reliability and Multivariate Analysis to Analyze Nursing Errors. Unpublished dissertation, Department of Industrial Engineering, North Carolina A&T State University, May 2009.
9. What Is Additive Manufacturing? Definition and Processes. (n.d.) Retrieved 1/15/2021 from https://www.twi-global.com/technical-knowledge/faqs/what-is-additive-manufa cturing
10. Schubert, Carl, van Langeveld, Mark C. and Donoso, Larry A. (2014). Innovations in 3D Printing; a 3D Overview from Optics to Organs. *British Medical Journal of Ophthalmology* 98(2, Feb), 159–161. Retrieved 3/1/2021 from https://pubmed.ncbi.nlm. nih.gov/24288392/
11. Vera, Alonso. (2010). What Is Human Factors-HF101. NASA. Retrieved 1/18/2019 from https://human-factors.arc.nasa.gov/web/humanfactors101/index.html#:~:text=Hu man%20Factors%20is%20an%20umbrella%20term%20for%20several,training%2C% 20both%20in%20flight%20and%20on%20the%20ground
12. What Is Human Factors and Ergonomics. (n.d.). Human Factors and Ergonomic Society. Retrieved 2/7/2021 from https://www.hfes.org/About-HFES/What-is-Human-Factors-and-Ergonomics
13. Anthropometry (n.d.). CDC, NIOSH. Retrieved 9/12/2021 from https://www.cdc.gov/ niosh/topics/anthropometry/default.html

14. Gaba, David M. (2000). Anaesthesiology as a Model for Patient Safety in Health Care. *BMJ* 320(7237, Mar 17). Retrieved 2/6/2021 from https://www.ncbi.nlm.nih.gov/pmc/articles/PMC1117775/

15. Al Khajeh, Ebrahim. (2018). Impact of Leadership Styles on Organizational Performance. *Journal of Human Resource Management*, 18, Article ID 678849. ISSN:2166-0018. https://ibimapublishing.com/articles/JHRMR/2018/939089/939089-1.pdfMicrosoftWord-939089-

16. Definition of Method. (n.d.) Dictionary. Retrieved 2/6/2021 from https://www.dictionary.com/browse/method

17. Trout, Johnathon. (n.d.). Process Improvement Explained. Retrieved 4//22/2021 from https://www.reliableplant.com/process-improvement-31857

18. Banton, Caroline. (2021). Just in Time. Corporate Finance and Accounting. Investopedia. Retrieved 2/22/2021 from https://www.iom/investopedia.com/terms/j/jit.asp

19. Hill, Kevin. (2018). Lean Manufacturing; the Seven Deadly Wastes. Retrieved 3/1/2021 from https://www.plantengineering.com/articles/lean-manufacturing-the-seven-deadly-wastes

20. Illustrated World of Proverbs. (n.d.). The Largest Collection of Proverbs in the World. Retrieved 11/24/2019 from www.worldofproverbs.com/2012/05/trust-but-verify-russian-proverb.htm

21. Working with Change Orders. (n.d.) JD Edwards EnterpriseOne Applications. Product Guide Data Management Implementation Guide. Retrieved 1/19/2021 from http:www.oracle.com/webapps/redirect/signon?=https:docs.oracle.com/cd/E / 16582-_01/doc/9le15128/work_w:ecos.htm

22. Koko, Maryam A., Burodo, Mohammed S. and Suleiman, Shamsudden. (2018). Queuing Theory and Its Application Analysis on Bus Services Using Single Server and Multiple Servers Model. Retrieved 2/27/2021 from http://www.sciencepublishinggroup.com/journal/paperinfo?journalid=104&doi=10.11648/j.ajomis.20180304

23. Kim-Brand, Donna. (2009) We Look but We Do Not See. *Ezine Articles*. Retrieved 2/22/2021 from https:ezinearticles.com/??We-look-but-don't-see=2534762

24. There Is a Way To Do It Better-Find It. (n.d.). Retrieved 4/15/2021 from: https://quoteinvestigator.com/2013/07/16/do-it-better/#:~:text=%E2%80%9CThere%E2%80%99s%20a%20way%20to%20do%20it%20better-find%20it%E2%80%9D.,the%20National%20Park%20Service%20used%20the%20expression%3A%208

25. Heathfield, Susan. (2021). How to Reduce Employees Resistance to Change. Ways to Reduce Resistance Before It Gets Started. 3+The Balance Careers. Retrieved 12/16/2021 from https://www.thebalancecareers.com/how-to-reduce-employee-resistance-to-change-1918992

26. Definition of Measurement. (n.d.). Measurements/Formulas-Types and Definitions. Retrieved 2/2/2021 from https://www.cuemath.com/measurement/#intro

27. Measurement Properties. (2007). American Thoracic Society. Retrieved 6/7/2021 from https://qol.thoracic.org/sections/measurement-properties/index.html

28. List of Confused Drug Names. (2019). Institute for Safe Medication Practices. Retrieved 9/16/20 from https:// www.ismp.org/recommendations/confused-drug-names-list

29. List of Error-Prone Abbreviations. (n.d.). Retrieved 2/11/2021 from https://www.ismp.org/recommendations/error-prone-abbreviations-list

30. Lesar, T.S., Briceland, L. and Stein, D.S. (1997). Factors Related to Errors in Medication Prescribing. *JAMA*, 277(4), 312–317. http://dx.doi.org/10.1001/jama.1997.03540280050033

31. Top 25 Lean Tools. Exploring Lean. (n.d.). Retrieved 2/18/2021 from https:www.leanproductio.com/top-25-tools.html

32. Elnaga, Amir and Imran, Amen. (2013). The Effect of Training on Employee Performance. *European Journal of Business and Management*, 3(4). ISSN 2222-2839 Retrieved 3/1/2021 from https:www.guide2research.com/research/training-industry-statistics

33. Neysetani, Behnam. (2014). Impact of Training on Employee's Performance and Productivity in the Construction Industry. Retrieved 1/18/2021 https://papers.ssrn.com/sol3/papers.cfm?abstract_id=2961057

34. Bouchrika, Inmed. (2020). 68 Training Industry Statistics:2020 Data, Trends and 3 Predictions. *Guide2Research.*

35. Purpose of Maintenance of Industrial Plant and Equipment, Reliability. Maintenance and Production Success. (n.d.). Retrieved 2/4/2021 from: https://lifetime-reliabilty.com

36. Chapter 5 Types of Maintenance Programs. (n.d.) O&M Best Practice Guide, Release 13. Retrieved 1/15/2021 from https://www1.eere.energy.gov/femp/pdfs/OM_5.pdf#:~:text=Chapter%205%20Types%20of%20Maintenance%20Programs%

37. Key Performance Indicators for Quality Assurance Examples. (n.d.). Retrieved 2/2/2021 from https://www.siteware.co/en/performance-management/kpis-,vquality-assurance-examples/:TOP 5 (siteware.co)

38. Agency for Healthcare Research and Quality. (2020). Quality Improvement and Monitoring at Your Fingertips. Retrieved 2/2/2021 from https://www.qualityindicators.ahrq.gov/Modules/iqi_resources.aspx

39. Generally Accepted Accounting Principles. (n.d.). Retrieved 3/11/2021 from https://www.bing.com/search?q=Generally+Accepted+Accounting+Principles+(GAAP)&cvid=d4115cdaecfc4fe08925e3aaaae942f5&aqs=edge..69i57j0l2.13224j0j4&FORM=ANAB01&PC=HCTS

40. Direct Labor- Definition, How to Measure. How To Calculate. (n.d.). Retrieved 11/29/2020 from https://corporatefinanceinstitute.com/resources/knowledge/accounting/direct-labor/

41. Work Sampling. (n.d.). Retrieved 1/21/2021 https://www.businessmanagementideas.com/production-management/techniques/work-sampling-objectives-theory-and-applications/7244\

42. Halton, Clay. (2020). Y2K. Economics, Macroeconomics. Retrieved 1/20/2020 from https://www.investopedia.com/terms/y/y2k.asp

43. Brisley, Chester L. (2001). *Sampling and Group Timing Technique. Maynard's Industrial Engineering Handbook*. 5th edition. McGraw Hill, New York. P. 17.47.

44. Zandi, Kjell, B. (2001). *Most Work Measurement Systems. H.B, Maynard and Company, Inc. Maynard's Industrial Engineering Handbook*. 5th edition. McGraw Hill, New York. P. 17.654.

45. Drucker, Peter. (n.d.). There Is Nothing so Useless as Doing Efficiently That Which Should Not Be Done at All. Retrieved 4/1/2021 from https://www.brainyquote.com/quotes/peter_drucker_105338?img=5

46. Assembly Line Balancing to Improve Productivity Using Work Sharing Method in Apparel Industry. *Global Journal of Research in Engineering. G Industrial Engineering*, 14(3). Retrieved 1/18/2021 from [PDF] Assembly Line Balancing to Improve Productivity using Work Sharing Method in Apparel Industry | Semantic Scholar.

47. Vaisnavi, N. (n.d.). Types of Incentive Plans-Wages-Human Resource Management. Retrieved 2/22/2021 from https://www.businessmanagementideas.com/wage/wage-incentive-plan/types-of-incentive-plans-wages-human-resource-management/12075

48. White Latex Adhesive. (n.d.). Retrieved 2/20/2021 from http://www.hhsj.chemchina.com/gzsjen/index.htm

49. Hill, Chase. (2019). These Tiny NASA Mistakes Had Huge Consequences. Retrieved 2/28/2021 from https://www.insidehook.com/article/science/tiny-nasa-mistakes-huge-consequences

50. Ressler, Kate. (2020). Bad Math; the Impact of Medication Dosage Miscalculations. Retrieved 3/1/2021 from https://www.pharmacytimes.com/

51. Illustrated World of Proverbs. (n.d.) The Largest Collection of Proverbs in the World. Retrieved 11/24/2019 from www.worldofproverbs.com/2012/05/trust-but-verify-russia n-proverb.htm

52. Proverbs 21.5 (n.d.). Retrieved 3/11/2021 from Benjamin Franklin stated "Haste makes waste."

53. Watson and Derouin. (2021). Critical Thinking: Learning from Mistakes and How to Prevent Them. Retrieved 3/28/2021 from https://www.amazon.com/Critical-Thinking-Learning-Mistakes-Prevent/dp/0367354608CriticalThinking:LearningfromMistake sandHowtoPreventThem:WatsonJr.,GeraldJ.:9780367354602:Amazon.com:

54. Robert, M. C., Choi, C. J., Shapiro, F. E., Urman, R. D. and Melki, S. (2018). Avoidance of Serious Medical Errors in Refractive Surgery Using a Custom Preoperative Checklist. *Journal of Cataract and Refractive Surgery*, 41(10), 2171–2178. Retrieved 11/11/2019 from ttps://www.sciencedirect.com/science/article/abs/pii/S0886335015011906

55. U.S. Department of Health and Human Services. (2006). Agency for Healthcare Research and Quality. Universal Protocol for Preventing Wrong Site. Wrong Procedure, Wrong Person Surgery. Retrieved 11/15/2019 from https://psnet.ahrq.gov/issue/univer sal-protocol-preventing-wrong-site-wrong-procedure-wrong-person-surgery#:~:text= The%20Universal%20Protocol%20applies%20to%20all%20accredited%20hospitals %2C,and%20decrease%20the%20overall%20risk%20of%20wrong-site%20surgery

56. Weingart, Gene. (2016). Wood Explorer. Retrieved 10/22/2019 from https://www.woo dworkingnetwork.com/wood/wood-explorer/sweetgum

57 What Are Direct Labor Costs. (n.d.). My Accounting Course. Retrieved 3/11/2021 from https://www.myaccountingcourse.com/accounting-dictionary/direct-labor-costs#:~:te xt=Definition%3A%20Direct%20labor%20costs%20are%20the%20wages%20or,dif ference%20between%20direct%20labor%20and%20direct%20labor%20costs

58. Thakur, Madhur. (n.d.). What Are Direct Labor Costs. Retrieved 4/1/2021 from https:www.wallstreet lingomojo/direct-labor-costs

59. Falls in the Workplace. (n.d.). NIOSH. CDC. Retrieved 2/22/2021 from https://ww w.cdc.gov/niosh/topics/falls/default.html#:~:text=Falls%20are%20a%20hazard%20 found%20in%20many%20work,affecting%20an%20ironworker%2080%20feet% 20above%20the%20ground

60. Hazardous Work. Occupational Safety and Health. (n.d.) Retrieved 2/2/2021 from http:// www.ilo.org/safework/areasofwork/hazardous-work/lang--en/index.htm

61. Ergonomics and Musculoskeletal Disorders. (n.d.). CDC, NIOSH. Retrieved 9/17/2020 from https://www.cdc.gov/niosh/topics/1}/nlecalc.html

62. Work Related Musculosketal Disorders and Ergonomics. (n.d.). Retrieved 3/1/2021 from https://www.cdc.gov/workplacehealthpromotion/health-strategies/musculoskele tal-disorders/index.html#:~:text=Work-Related%20Musculoskeletal%20Disorders%20 %26%20Ergonomics%201%20Carpal%20tunnel,and%20Implementing%20Workplace %20Controls.%20...%205%20Ergonomics.%20

63. Training Requirements in OSHA Standards. (n.d.). Retrieved 4/15/2021 from https:// www.osha.gov/sites/default/files/publications/osha2254.pdf

64. Kwok, Angus C. F. (2014). The Evolution of Management Theories: A Literature Review. Retrieved 08/25/2019 from http://www.ny.edu.hk/web/cht/nang_yan_busines s_journal/Nang%20Yan%20Business%20Journal/Kwok,%20A.%20C.%20F.,%202014. %20The%20Evolution%20of%20Management%20Theories%20-%20A%20Literature %20Review.pdf

65. Taylor, Francis Winslow. (1911). The Principles of Scientific Management. Retrieved 11/2019 from https.pearsoncustom.com/wps/media/objects/2429/18.2487430/pdfs/taylor.pdf

66. Price, Brian. (1989). Frank and Lillian Gilbreth and the Manufacture and Marketing of Motion Study, 1908–1924. Retrieved 8/22/2021 from http://web.mit.edu/allanmc/www/TheGilbreths.pdf

67. Giges, Nancy. (2012). Lillian Moller Gilbreath Biography. *ASME. Org.* Retrieved 10/6/2019 from https://www.asme.org/topics-resources/content/lillian-moller-gilbreth

68. Gilbreth, Lillian Moller. (n.d.). Mother of Modern Management. Retrieved 9/19/2019 from https://www.sdsc.edu/ScienceWomen/gilbreth.html

69. Sullivan, Jane. (2010). Incorporate Gantt Theory into Your Work Schedules to Boost Productivity. Retrieved 9/19/2019 from https://www.business.com/articles/management-theory-of-henry-gantt/

Index

M

Maintenance, xviii, 4, 10, 13, 35–39, 64, 70, 80, 85, 93, 98, 99, 101, 103–105, 110, 115, 118, 121, 137, 148, 152, 175
Method, xix, xxv, 5, 7, 9, 11, 14, 23, 24, 26, 32, 36, 37, 42–45, 52, 60, 71, 74, 80, 98, 100, 101, 106, 115, 116, 120, 124, 126, 140, 158, 159, 163, 171–174
Methods improvement, 4, 32, 34, 36, 42, 72

N

Non-value added, 81

O

Operator rating factor, 48, 52, 79, 84, 92
OSHA, 11, 53, 63, 114, 115, 158, 159, 161, 164, 167
Overhead, xix, 3, 7, 8, 14, 35, 42, 58, 64, 96, 137, 153, 154, 164

P

Packaging costs, xiv, 3, 7, 42, 153, 187, 188
Predetermined time standards, 45
Productivity, xiv, xvi, xvii, xviii, 1, 3, 9, 10, 14, 21, 22, 31–33, 36, 37, 39, 42, 46, 48, 49, 51, 53, 56, 61, 64, 65, 67–69, 71, 77–79, 81, 84, 91, 99–101, 116, 161

R

Raw materials inventory, 136

S

Safety team, xxiii, 36, 93, 157–159, 161, 163, 164, 167–169
Six Sigma, 7
Standard direct labor hours, 42, 47, 67, 68, 131, 153
Standard hours, xxv, 3, 24, 42, 47, 48, 50, 52–54, 61, 62, 71, 74, 77, 78

T

Taylor, Frederick, 189
Therblig, 45, 190

U

Union, 10, 52, 68, 70, 72, 73, 79, 85, 93

W

Work in process, xiv, xviii, 7, 9, 12, 14, 24, 29, 35, 42, 45, 74, 89, 90, 109, 120
Work in process inventory, xviii, 9, 14, 24, 29, 35, 74, 109, 120, 173

For Product Safety Concerns and Information please contact our EU
representative GPSR@taylorandfrancis.com
Taylor & Francis Verlag GmbH, Kaufingerstraße 24, 80331 München, Germany